ANTIOXIDANTS

Natural and Synthetic

Dr. Asim Kumar Mukhopadhyay

Amani International Publishers

Kiel 2006

Bibliographic information published by the Deutsche Nationalbibliothek:

The Deutsche Nationalbibliothek lists this publication in the Deutsche Nationalbibliografie; detailed bibliographic data are available in the Internet at http://dnb.d-nb.de.

Manufactured in Germany

E-Mail: service@amani-publishers.com

http://www.amani-publishers.com

ISBN-13: 978-3-938054-05-5

ISBN-10: 3-938054-05-0

Dedicated to my mother whose single-minded devotion and inspirations acted as a catalyst for my development as a professional and a scientist of no mean achievement.

PREFACE

It is a paradox that on one hand oxygen saves hundreds of critical patients by rejuvenating their lives, and has multiple users in industry including steel plants, on the other oxygen is the primary reason for oxidative degradation of foods and rancidity of fats and oils and polymers including plastics, rubber and also paints.

In other words oxygen surprisingly acts as a poison and a degrading agent – an inhibitor – for deterioration of food and related products particularly those in packed condition. To annul the negative effect of Oxygen, Nitrogen blanketing has been effectively used as an inhibitor of oxidative degeneration. At least this helps in increase of shelf life and protects the packaged materials for a longer period. It may however be added that some foods including fats and oils have inherent oxygen deterrents which neutralizes effect of oxygen for quite some time. Food and other products are also saved from aerobic bacteria and fungi due to elimination of oxygen or presence of minimal quantity of oxygen.

In this context anti-oxidants – a group of several organic substances including certain vitamins and minerals – have been found to counteract the damaging but otherwise normal, effects of the physiological process of oxidation. These antioxidants are also believed to play a role in preventing the development of a number of chromic diseases as cancer, heart disease, stroke etc.

Antioxidant chemistry has developed rapidly. In addition to antioxidants present in natural substances, scientists have developed a number of synthetic antioxidants, which have either replaced or supplemented natural antioxidants. Ironically antioxidants are not always effective or safe for a long period as they themselves are oxidized and their utility diminishes with time. Nevertheless antioxidants have effective uses for a reasonable time.

For many a decade Butylated hydroxy toluene, also known as di-isobutyl para cresol, abbreviated as BHT or DBPC, was the most prominent antioxidant having uses in diversified disciplines of food products, polymers, fats and oils. BHT was considered very safe for both human food and animal feeds (Cattle, Poultry etc.). However, things changed with time. Recently scientists have found out that BHT has carcinogenic effects on human foods and has been banned for such uses. However, in many developing countries BHT is still used in food products because of its easy availability and relative cost advantages.

Traditionally BHT has been used in petroleum industry in lubricating oils, as a fuel additive, in petrochemicals, particularly polymers and even in vegetable oils such as palm oil. BHT still remains as the most popular antioxidant in diverse fields.

USA, Western Europe and Japan had availability of both the critical raw materials such as para cresol and isobutylene. Collectively they were producing 80000 – 100,000 tpa BHT for various uses. Other countries such as China, India, and South Korea were also producing BHT for diversified uses.

For very many years I had intimate contacts with many manufacturers of BHT particularly in India. This was because my company, Atul Products Limited, where I was the Chief Executive for many years, was the country's only producer of mixed cresols from Toluene feed stock. Based on in-house R&D work we were successful in isolating p-cresol (95%) from mixed cresols and thereafter 98-99% pure p-cresol via crystallization and centrifuging. Mother liquor from the crystalizers itself had uses as an antioxidant for some petroleum products. Being the only manufacturer of para cresol we were supplying the material to all Indian manufacturers of BHT. Some quantities were even exported.

Once the importance of antioxidants was realised, a number of antioxidant research centers and laboratories were established in USA, Western Europe and Japan. The importance of some natural antioxidants such as Vitamins C, E and A was understood and soon natural Vitamins were also substituted and supplemented by synthetic antioxidants. BHT which was not any longer recommended for human consumption was substituted by other antioxidants such as Butylated hydroxy anisole (BHA) and tert-btuyl- hydroquinone (TBHQ) which were considered safe for prevention of oxidation for human food products With some modifications BHA and TBHQ could be produced in BHT plants. However, it must be emphasized that BHT is sill the single largest volume antioxidant, and, other butylated products including butylated hindered phenols are not only more costly but are still in limited use.

In India in our own plant facilities and R&D laboratories, at Atul we successfully developed BHA and TBHQ in conjunction with the existing manufacturers of these antioxidants.

On my part I was not only interested in Butylated phenol/ cresols and polyphenols but also with the natural network antioxidant. This will be discussed in some details in the later chapters.

I am of the considered opinion that this book will be of immense value to the students, professors, practicing professionals and others. Industry in general including users of antioxidants and manufacturers will find this as an extremely useful reference book.

THE AUTHOR

Dr. Asim Kumar Mukhopadhyay, the author of this book, a highly qualified techno-commercial professional, had his earlier education in the district town of Bankura, West Bengal, India. He graduated in Chemical Engineering from Jadavpur University, Calcutta and earned his Master's degree in Chemical Engineering from Indian Institute of Technology, Kharagpur. He received his Ph.D. in Chemical Technology from Moscow Institute of Petrochemicals and Gas Industries, Moscow, then the capital of U.S.S.R.

After completion of his studies Dr. Mukhopadhyay joined the industry and became a multidisciplinary professional by handling different portfolios encompassing process and project Management, R&D Management, Production and Marketing Management, Financial Management, etc.

Dr. Mukhopadhyay held top positions in organizations of repute. He was the first Managing Director of WEBCON, a multi-functional development consultancy organization promoted by Industrial Development Bank of India (IDBI) and other financial and promotional institutions and Banks.

He was whole time Director of IVP Ltd. Bombay for four years and catalysed the growth of the company. For ten and half years he was the chief executive officer of Gujarat Aromatics Ltd., later rechristened as the Aromatics Division of Atul Products Ltd., Valsad, Gujarat State and was instrumental in converting this single product sick unit to a highly profitable multiproduct establishment and a leader in the field of cresols and downstream derivatives. Since his early separation from Atul, Dr. Mukhopadhyay has been working as a national and international consultant and helping in the development of cresols and downstream derivatives including fine chemicals and antioxidants.

A widely travelled person, Dr. Mukhopadhyay is a versatile writer having authored several technical articles, short stories, poems and two books. He is truly a multifaceted personality.

ACKNOWLEDGEMENT

It is with immense pleasure that I deeply appreciate the assistance, direct or indirect, received to fulfill my desire to write a book on antioxidants.

I take this opportunity to express my deep sense of gratitude to the chief librarians of Harris County Public Library at Katy near Houston, Texas, USA and the Indian Institute of Technology, Mumbai, India for using their library facilities.

My sincere thanks are due to the authors and publishers of various books, periodicals and technical articles in the allied fields.

I must keep on record my deep sense of gratitude to a number of experienced and knowledgeable officials of a few global companies such as Sumitomo Corporation Japan, Sumitomo Chemicals Co. Ltd., Japan, Rhone- Poulenc, USA, Merisol Co. USA and South Africa, who had been active in the fields of cresols and derivatives including synthetic antioxidants. I must express my sincere appreciation and gratitude to these people who helped me in my global research on the antioxidants related business.

Finally, I express my deep sense of gratitude to my wife, Mamata, for her positive support and assistance that catalyzed me to write this book.

CONTENTS

Page No.

Contents (contd.) Page No.

Contents (contd.) Page No.

Contents (contd.) Page No.

1

INTRODUCTION

1.1 WHAT ARE ANTIOXIDANTS?

Antioxidants are essentially a group of organic chemicals including Vitamins (E, C and A), (minerals such as selenium and occasionally zinc and Manganese as micronutrients), and carotenoids, the coloring materials that include β-carotein, α-carotene, β-cryptoxanthin etc. and also phytochemicals. These are being discussed in some details in the subsequent chapters.

Antioxidants have been used to retard the process of oxidative degradation of food products, polymers like rubber and plastics, fats and oils and others such as elastomers, fibers, fuels and lubricating oils. Even in the process of aging oxidation plays a part in humans and other animals.

Oxidative degeneration has been stated to have been caused by harmful molecules called free radicals. Free radicals are molecular fragments having one or more unpaired electrons, usually short lived and highly reactive following the usual mechanism of chain reaction of initiation, propagation and termination. A free radical is conventionally indicated by a dot, as in Cl', $(C_2H_5)'$ etc. In spite of their transitory existence, they are capable of initiating many kinds of chemical reactions by means of a chain reaction. Free radicals are formed only if 1) radicals attack the substrate and 2) the radicals lost by this reaction is regenerated [1]. Free radicals promote beneficial oxidation that produces energy and kills bacterial invaders. In excess, however, they cause harmful oxidation that can damage cell membranes and cell contents.

Examples of free radicals are the super oxide anion, hydroxyl radical, transition metas such as iron and copper, nitric acid and ozone (O_3). Free radicals containing oxygen (O_2) known as reactive oxygen species (ROS), are the most biologically significant free radicals. Also included are derivatives of oxygen that do not contain unpaired electrons, such as hydrogen peroxide (H_2O_2) singlet oxygen and hypochlorous acid.

Nitric oxide, NO, CAS No. [10102-43-9] having molecular wt. of 30.01 is also a free radical produced by many different cells of the immune system. Nitric oxide has many important roles in the body:

- it is an important signalling molecule that turns genes on and off,
- by controlling the muscular tone of blood vessels, it regulates circulation and normalizes blood flow,
- it modulates communication between brain cells and maintains rejuvenation of the memory system.

However, Nitric oxide can be very destructive otherwise:

- it restricts blood flow and contributes to heart disease and stroke,
- if in excess, Nitric oxide can catalyse diseases like arthritis, bowl disease and even cancer,
- it promotes more free radicals and this can be dangerous for the system,
- it may cause memory loss and brain aging [2].

1.2 Natural Antioxidants

In addition to enzymes, some vitamins and some minerals mentioned earlier having proven antioxidation properties, there are other compounds known as phytochemicals and Flavonoids, though still controversial, have known antioxidant properties. Even uric acid a product of DNA metabolism has been recognized as an important antioxidant.

Out of many antioxidants mentioned earlier, vitamins C and E, glutathione, lipoic acid and coenzyme Q_{10} (C_oQ_{10}) have been classified as very prominent and rightly recognized as network antioxidants. This will be discussed in subsequent chapters. [2]

Vitamins E and C and Ginko biloba are three of the most important and natural antioxidants consumed in the world. This will call for some detailed discussions at the appropriate place to throw more light and enlighten readers.

1.3 Synthetic Antioxidants

To supplement, substitute and augment the antioxidants, which were earlier obtained only from natural sources, a number of synthetic antioxidants were introduced a few decades ago. These synthetic antioxidants, mostly hindered and butylated phenols and polyphenols, have proved to be cheaper than many natural antioxidants and quite effective in the retardation process of oxidative degeneration. Butylated hydroxy toluene (BHT), largest volume antioxidant made from easily available raw material such as paracresol and isobutylene, used to be made in substantial quantities in USA, Western Europe, Japan and elsewhere. BHT has been the most attractive antioxidant for petroleum products including petrochemicals. Because of somewhat carcinogenic properties, use of BHT has been restricted for human food products. Still it remains as the most popular multi-use antioxidant in the world. 80,000 to 100,000 MTPA of BHT was produced collectively by USA, UK, Germany and Japan during late eighties and early nineties. Apart from petroleum products BHT has been commercially used for cattle and poultry feed products.

India and China, two of the fastest developing countries also have expanding petroleum refining and petrochemical industries. Accordingly good quantities of BHT are consumed by China and India and also by other countries such as South Korea, east European countries South Africa, Australia, some countries in south-east Asia and elsewhere. Considering the world as a whole use of BHT is not declining but still on the rise. Its price is in the vicinity of $1.6 – 2 per kg and that makes it by far the cheapest antioxidant. Apart from BHT two other prominent and promising antioxidants have been Butylated hydroxy anisole (BHA) and tert-butyl-hydroquinone (TBHQ). Other butylated phenols, cresols and xylenols have been also used in some quantities particularly by the petroleum industry last 10-15 years. Irganox^{TM}-group of antioxidants introduced by Ciba-Geigy have proved to be very effective. Now that patent rights have expired a few more companies have started or planning to start making antioxidants of Irganox type.

1.4 Less Known Important Antioxidants

Ginko biloba, a strong, hardy tree, is of Stone Age origin and is one of the oldest trees on the planet. It was destroyed during the ice age everywhere except in China, where it still lives and thrives. It was brought to USA in 1784 and is now a common tree in the states.

The leaves and fruit of the Gingko have been used as a medicinal treatment in China since 2000 B.C. Its antioxidant value was well recognized in China and effectively used primarily for problems related to brain function, heart disease and other circulatory disorders. Gingko biloba, a flavonoid, has been best known as the antioxidant for human brain. Gingko today is one of the top three natural antioxidants along with Vitamin E and Vitamin C.

Glutathione, CAS No. [70-18-8] [4] a major antioxidant along with lipoic acid and Coenzyme (Q_{10} / C_oQ_{10}) are produced by the body but levels of these antioxidants decline as we age and it is necessary to supplement all of them at the proper time. Glutathione is produced by the body from three amino acids found in food. [2]

Coenzyme Q_{10}(C_oQ_{10}) is an effective treatment for heart failure, high blood pressure and might have uses for cancer.

Relatively new to the United States, lipoic acid has been used in Europe safely for last 30-40 years in treatment of diabetes. More about these relatively unknown antioxidants in a subsequent chapter.

1.5 Synergistic Effects

The synergistic effect of antioxidants in combination has been well established in the food industry. To give an example BHA in combination with Propyl gallate, CAS No. [121-79-9], [4] is a very powerful antioxidant. Citric acid, Cas No. [77-92-9], a well-known food additive-cum-stabilizer in combination with BHT/BHA becomes a very powerful antioxidant.

It needs to be emphasized that an antioxidant has certain features that makes it acceptable for food use. For food uses for many years it was considered safe to use natural antioxidants. In more recent years, however, some of the synthetic compounds were favored because of their excellent antioxidant properties. In the USA uses of food additives /antioxidant must strictly conform to the laws in force such as Federal food, Drug and cosmetic act, Federal meat inspection, and/or Federal poultry inspection act [3]. In many countries similar laws have been enacted and are strictly compiled with.

1.6 Antioxidant Classes (Synthetic)

Based on various considerations antioxidants have been classified into different groups either from structural similarities or functionality. Some suggested classes are hindered phenols, Aromatic amines, Organo sulfur compounds, Phosphorous compounds, hindered amines etc. (For details one may refer to Ullman's encyclopedia of industrial chemistry Vol.3, pages 634-642):

- Hindered phenols are a major class of antioxidants, which have been used for oxidative degradation of various organic products such as plastics, elastomers, synthetic fibers, fuels, lubricants and foods. Stearic hindrance is provided by effective substituents on the ortho position in the ring with respect to hydroxyl group. 2,6-Di-tert butyl phenol, 2,6-di-tert-butyl-4-methyl phenol (BHT) etc. are some of the multi use hindered phenols. Some of the prominent hindered phenols and cresols, are elaborated in Chapter 3. Butylated phenol, mono and di-butylated para- and meta cresols, Shell's Ionox^{TM} hindered phenols and Ciba-Geigy's trade mark Irganox group of hindered phenols are very prominent products which have been effectively used in various diverse applications. Chapter 3 discusses briefly about these products.

- Secondary aromatic amines primarily are used in elastomers stabilization. Mainly they have been grouped as diarylamines, amine-ketone condensation products (such as aniline-acetone condensates) and substituted p-phenylene diamines. Most of them are colored and form highly colored oxidation products. Carbon black-filled, vulcanized elastomers are a major application. For them discoloration is not a drawback.
- Organo-Sulfur compounds as antioxidant stabilizers had been in use for quite sometime. Metal salts or dialkyl dithiocarbonic acids, Zinc dialkyl dithiophosphates and esters of thiodipropionic acid have been important commercial antioxidants in the diverse fields of lubricants, elastomers and plastics.
- Trivalent phosphorous compounds i.e., phosphites, and phosphonites are perhaps the only phosphorous compounds used as antioxidants in industry. Only other major class is zinc dialkyldithiophosphate, which is used in lubricants.
- Hindered amines are used primarily as light stabilizers but were also used as antioxidant. Stabilizers against thermo-oxidative degradation.

  In certain instances, at temperatures less than 120°C, some hindered amines are more effective than hindered phenols for stabilization of polypropylene stretched tapes.

Specialized practices have evolved for each of the following major substrate classes that are stabilized with antioxidants.

1. Foods
2. Fuels
3. Lubricants
4. Polymers

Major antioxidants for retardation of oxidative degradation of the above products are being specified and discussed in appropriate places.

1.7 Natural Antioxidants: Importance of Recovery

Natural antioxidants (food antioxidants) are structurally phenolic or polyphenolic compounds commonly occurring in plant materials, which interfere with the formation of free radicals (the initiation reaction) and also deter the propagation of oxidation in the free radicals chain. This prevents the formation of peroxides.

It is well known that some antioxidants occur naturally in all foods and herbal medicines. However, their addition in small quantities, within permissible limits, delay, retard or prevent development of rancidity caused by atmospheric oxygen and thus preserve fats and oils, and fat soluble components like some vitamins, carotenoids and other nutritive ingredients of food.

Dietary supplementation of natural antioxidant nutrients such as Vitamins A, C and E and flavonoids to foods may prevent many human diseases caused by oxidative damage including aging, cataracts, coronary heart diseases, Cancer etc. Their natural occurrence in foods, such as, fruits, vegetables, leaves, nuts and grains provide protection against oxidative attack. Tremendous benefit will be achieved if such diseases could be delayed by a few years.

It is to be noted, however, that when food commodities are subjected to processing or during their preservation and storage, antioxidants present in food often get depleted. Consequently such foods are fortified with additional quantities of antioxidants either from natural sources or synthetic antioxidants.

Synthetic antioxidants are cheap though natural antioxidants are also attractive as additives in food items. The superiority of natural antioxidants has been proven vis-a-vis synthetic ones in terms of safety, tolerance and non-toxicity. In general, synthetic food antioxidants are subjected to pharmacological and technical scrutiny and evaluation for mutagenic, carcinogenic and pathogenic effects. After all consumption of food items is critical so far as human healths are concerned.

Most commonly used natural antioxidants are Vitamin C (ascorbic acid), Vitamin E (tocopherols), citric acid, flavonoids and others. Broadly there are five major types of antioxidants:

1. Primary antioxidants are those compounds, namely phenolic substances, that terminate the free radical chains in lipid oxidation and function as electron donors, e.g., natural and synthetic tocopherols, alkyl gallates, BHT, BHA, TBHQ etc.
2. Oxygen scavengers, substances that react with oxygen and can thus remove it in a closed system, e.g., ascorbic acid (Vitamin C), ascorbyl palmitate, erythorbic acid (iso-ascorbic acid) etc.
3. Secondary antioxidants, those compounds, which function by decomposing the lipid hydroperoxides into, stable end products.
4. Enzymic antioxidants, which function either by removing dissolved or head space oxygen, e.g., glucose oxidase, or by removing highly oxidative species, e.g., super oxide dismutase.
5. Chelating agents, syngerstic substances that enhance the action of phenolic antioxidants, e.g., citric acid, amino acid etc.

1.8 Spice and Herbal Extracts as Antioxidants

Spices and herbs are used not only to enhance flavor but also shelf life of various foods. It was demonstrated that 32 spices could behave as antioxidants out of which rosemary and sage belonging to the Labiate family were identified as the most potent antioxidants. Clove, mace, savory, oregano, nutmerg, turmeric, ginger, thyme etc., also possess significant antioxidant properties. These have been discussed in some details by Mukhopadhyay M. vide her book "Natural Extracts using Supercritical Carbon Dioxide, CRC Press LLC, Florida, 2000".

1.9 Plant Leaf Extracts as Antioxidants

Plant leaves like any dark green leafy vegetables are rich sources of phenolic compounds, which are lipid soluble antioxidants. These include α-tocopherol (Vitamin E) and β-Carotene (Provitamin A) as chlorophyll and carotenes are closely related.

The antioxidant activity of several leaf extracts of mediterranean plants has been reported. α-tocopherol was identified as the main antioxidant in leaf extracts of many plant species. The tocopherol content could be enriched by hexane extraction upto 4.7% from the dry leaves. The highest tocopherol content was earlier reported to be as high as 1000 ppm in the leaves of Vritica dioica. Such high level of tocopherol is even more than what most common vegetable oils contain and is much higher than the leaves, both edible and non-edible types. This suggests that these leaves are even much better sources of Vitamin A and Vitamin E, then the seeds. (Reference same as in 1.8).

It is foregone conclusion that antioxidants have to be added in order to increase the storage life of food products. This would prevent food spoilage for a reasonable period. Simultaneously due attention has to be paid for quality of the antioxidant to avoid rancidity and toxicity that might develop after use of such antioxidants. The final product for human consumption should not be carcinogenic. A classical example is BHT, which earlier considered as safe (GRAS) as a food antioxidant has now been prohibited as a food additive in most of the countries.

All aspects of both natural and synthetic antioxidants will now be examined in depth in the following chapters.

2

NATURAL ANTIOXIDANTS

2.1 ANTIOXIDANTS FROM PLANT AND ANIMAL SOURCES

During the last one hundred years the importance of use of antioxidants for retardation of oxidative degeneration of food products during processing, preservation and storage for a reasonable period was well recognised. Various antioxidants laboratories and research centers throughout the world were established to study the effectiveness of use of different antioxidants and effects on toxicity and carcinogenic properties after addition of antioxidants. This was more apparent for use of new and hitherto unknown antioxidants considering various aspects including safety of human lives after consumption of antioxidant-added food products. For introduction and commercialization of new food antioxidants this was of paramount interest and importance.

A series of antioxidants from natural sources both plants and animals, were extracted and isolated using different technologies for uses including those for saving human lives or for preventing premature aging. Some scientists preferred to call this as antioxidant miracle. One classical example is Packer Laboratory in University of California (Berkley). Scientists there were successfully working on many known and also for quite a few new antioxidants. For the human systems they even concluded that five antioxidants such as Vitamin E, Vitamin C, Glutathione, Lipoic acid and coenzyme (C_OQ_{10}) could be best described as network antioxidants and extremely useful for human lives. It was also found that Vitamins C and E were not produced in the human bodies and should be obtained from different food products. Other three antioxidants are produced by the human body at inadequate levels particularly during the process of aging. It was thus rightly concluded that all these antioxidants are to be supplemented. [2]

Then came the petroleum era. Petroleum products, both fuels and lubricating oils should be protected against deterioration and increase of shelf life by incorporating suitable antioxidants. Even petrochemicals including polymers and plastics needed addition of suitable antioxidants. These are mostly available from synthetic sources, which will be discussed in subsequent chapters. Even Fats and Oils such as lard and palm oil needs specific antioxidants during storage. Again these are mostly synthetic antioxidants. Natural antioxidants, though effective, will be very costly and may not be commercially attractive for such applications.

Most of the important natural antioxidants elaborating their chemistry, properties, uses etc. are discussed in this chapter. Isolation of antioxidants from natural extracts will be highlighted. Oxidation of natural antioxidants will be discussed in some details.

2.2 Various Natural Antioxidants

Some of the natural antioxidant groups or individuals, which have been studied in great details and have become popular and extensively used in food products, are shown in Fig. 2.1.

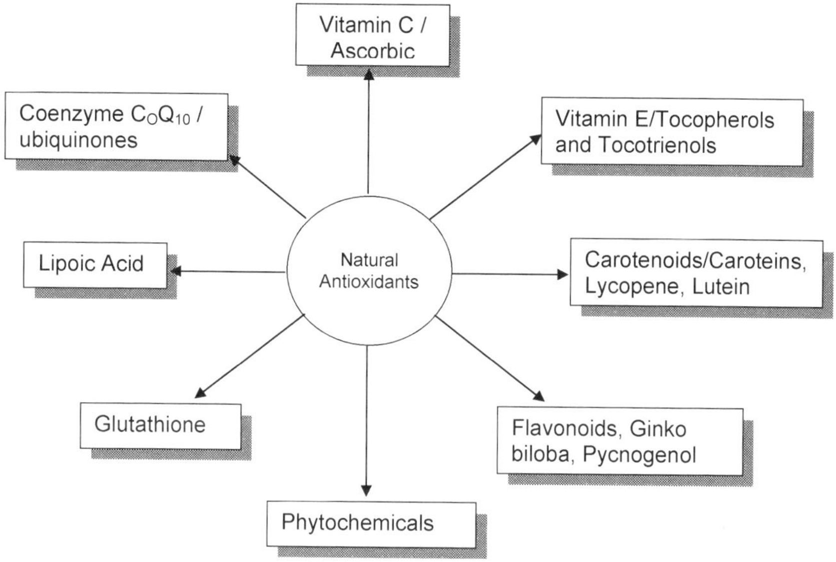

Fig. 2.1 Various natural antioxidants

In April 2000, the Food and Nutrition Board of the Institute of Medicine, an advisory group that is part of the National Academy of Sciences in USA reported that Vitamins C and E, Carotenoids like β-carotene, Selenium etc., should come from food products only and need not be supplemented. For each type of nutrient, the Board has established a Recommended Dietary Allowance (RDA) and a tolerable upper intake level (UL) – the maximum amount of nutrient that healthy individuals can take daily without affecting their healths.

Since then the Board has been updating and expanding the system for determining RDA and UL values. Table 2.1 summarizes these values recommended by the 2000 report of the Board: [5]

Table 2.1 The food and Nutrition Board's recommendations for uses of some antioxidants

Antioxidant	RDA (adults)	UL (adults)	Comments
Vitamin E	15 mg	1070 mg (natural) 785 mg synthetic	Higher amounts impair blood clotting, increasing likelihood of hemorrhage
Vitamin C	Women 75mg, Men 90 mg	2000 mg	Higher levels could lead to Diarrhea, very high levels may lead to cancer, kidney stones etc.
β-Carotene	None	None	Chromic high doses not advisable
Selenium	55 micrograms	400 micrograms	Higher amounts could create a number of problems such as hair loss

Similar Boards or governing bodies are existing in other countries such as UK, Germany, Japan, China, other European Countries, India etc. But recommendations of the US Food and Nutrition Board might set examples to other countries, since human hearts are concerned. Safety regulations are to be strictly adhered to.

2.3 Sources of Antioxidants in Selected Food Products

As a result of detailed studies for locating antioxidants in various food products it was demonstrated that some of the fruits and vegetables and a few animals are rich sources of a few antioxidants.

- Vitamin E is found in vegetable oils, walnuts, peanuts almonds, seeds, olives, wheat germ, and leafy green vegetable.
- Vitamin C rich vegetables and fruits include citrous fruits (such as oranges, grapefruit, amla etc.), straw berries, broccoli, leafy green vegetables, tomatoes, peppers, potatoes, strawberries and others.

- Common sources of β-Carotene (Provitamin A) are mainly: carrots, mangoes, papaya, pumpkin, spinach, sweet potatoes, peppers, apricots, etc.
- Selenium as essential micronutrient with antioxidant properties is found in seafood, beef, pork, chicken, brown rice and whole-wheat brad. [5]

For food products it was always considered safe to use natural antioxidants. An antioxidant for use in food products has certain features that make it acceptable for food use:

i) An antioxidant should not impart undue flavor to the system

ii) It should be convenient to use

iii) It has to be safe for handling

iv) It should be low in cost

v) It should be effective in low concentrations.

In the United States antioxidant uses are strictly in accordance with

- Federal Food, Drug and Cosmetics act
- Federal meat inspection act
- Federal poultry inspection act

Similar acts and laws are in practice in most of the developed and developing countries. A type of 'bird flu' was recently noticed in many of the Far East countries particularly China, Thailand and elsewhere. Chickens affected by this disease had badly affected human lives also. Accordingly thousands of chickens were killed so that the disease does not spread. Though antioxidants in feed products might not have anything to do with this disease. But it shows that consumption of poultry products (meat and egg) should be always monitored and due attention has to be paid so that human lives are not affected. This is precisely the reason why stringent laws have been enacted in most of the countries. This is also true for use of antioxidants in cattle and poultry feed.

Very stringent regulations have been enforced in USA and in most European countries and also in developing countries like China, Thailand, India and elsewhere. In most of these countries types and maximum quantities of antioxidants permitted in fats, fat containing foods and packaging materials for processed foods. One of the desirable characteristics of an antioxidant is its ability to, 'carry through' the heat treatment process. Gum guaiac, an antioxidant, for example, has passed through this test.

An enzyme system known as glucose oxidase catalase is listed as an antioxidant because it removes oxygen from hermetically sealed containers. By oxygen removal, the deteriorative reaction due to oxygen is eliminated. Should be realized that the system of preventing oxidation and off-flavors is different from other chemicals listed as antioxidants. For a specific application, it is necessary to identify the approved antioxidant for food products while achieving effective results economically. [6]

2.4 Some Aspects of Antioxidants from Natural Sources

Mother nature has already incorporated different types of antioxidants in plants and animals. Some of them are unique in the sense that they could not be yet synthetically prepared. Attempts are however being made by several laboratories to reproduce and synthesize many of these antioxidants. Natural antioxidants encompassing a brief chemistry, properties etc. are briefly discussed here.

2.4.1 Glutathione [4]

CAS No. [70-18-8]

mol. formula	$C_{10}H_{17}N_3O_6S$
mol. wt.	307.33

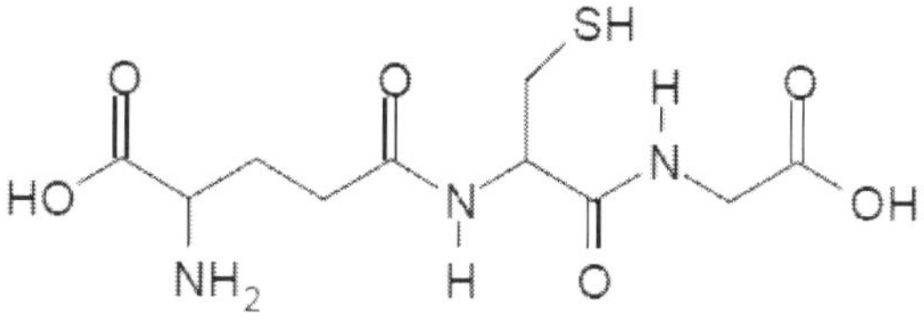

Crystals from 50% alcohol, mp 195°C, freely soluble in water, dil. alcohol, liquid ammonia dimethylformamide. It is a major low mol.wt. thiol component (containing S) of the living plant or animal cell. It has been isolated from yeast. [4]

Glutathione has been termed as Nature's master antioxidant. It has been stated that low levels of glutathione may cause illness and premature death. It is produced in the body. It is instrumental in the detoxification of drugs and pollutants and for healthy liver function. Glutathione is important for strong immune system. It is also stated that boosting glutathione can possibly reverse age-related retardation of immune system.

Glutathione is abundant in fruits, vegetables and freshly cooked meat. However, it is supposed to break down during digestion. [2] Glutathione rejuvenates the immune function of the body and is supposed to be an anti-aging antioxidant. Glutathione, chemically also known as γ-glutamyl cysteinyl glycine, is a universal component of the living cell, contains glutamic acid, cysteine and glycine. These are chemically bound but can be separated by hydrolysis.

Glutathione has been essentially used as high value nutrition component and is also used in the field of metabolic research. [1]

2.4.2 Lipoic Acid

Also known as 6,8-dithio octanoic acid, thioctic acid.

CAS No. [62-46-4]

Chemical Structure	SS CH_2 CH_2 CH $(CH_2)_4$ COOH
mol. formula	$C_8 H_{14} O_2 S_2$
mol. wt.	206 132
	Crystals
mp	60 – 61° C
bp	160 – 165° C

Practically insoluble in water, soluble in fat solvents, forms a water soluble sodium salt.

Lipoic Acid is mostly available from yeast and liver. This is a pyruvate oxidation factor. Pyruvate is a normal intermediate in carbohydrate metabolism. [1]

Lipoic acid has been classified as a universal antioxidant. This offers powerful protection against three common ailments of aging: stroke, heart disease, and cataracts. Lipoic acid strengthens memory and prevents aging. It functions as a booster for other antioxidants such as Vitamins E and C, glutathione and coenzyme Q_{10}. Surprisingly lipoic acid is new to the United States though it has been safety and effectively used in Europe for almost 25 years to prevent and relieve the complications of diabetics. Apparently lipoic acid turns off bad genes that can accelerate aging and cause cancer. It has been successfully used to treat other liver diseases such as hepatitis C. Reportedly it is synthesized by the body and is also present in small quantities in potatoes, spinach, and red meat. [2]

Lipoic acid, a versatile super antioxidant, practically does everything an antioxidant should do. Lipoic acid is produced by animals, humans and even plants but in insignificant quantities. A lipoic acid supplement has been recommended for meeting the demand for human bodies, since sufficient amounts would not be available from food products.

2.4.3 Ubiquinones, Coenzyme (C_OQ_{10})

CAS No. [303-98-0]

ubiquinones group formula is $CH_3C_6O_2(OCH_3)_2$ [CH_2OH: $C(CH_3)$ $CH_2]_n$ H

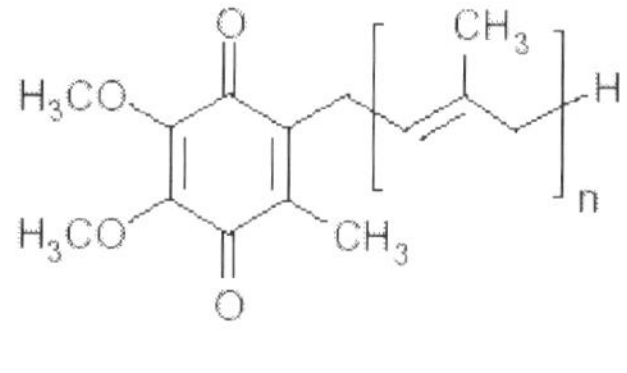

According to the dual system of nomenclature the compounds can be described as Q_n in which n = 1-12 or Ubiquinone (x) in which x designates the total number of carbon atoms in the side chain and can be any multiple of 5. Naturally occurring members are the coenzymes Q_6-Q_{10}. Ubiquinone Q_{10}(C_OQ_{10}), an active enzyme is an antioxidant of great interest. They are active in citric acid cycle in carbohydrate metabolism. [1, 4]

The entire series has been prepared synthetically. These coenzymes are also found in the majority of aerobic organisms from bacteria to higher plants and animals. These are comparatively low molecular wet. organic substances that can attach itself to, and supplement a specific protein to form an active enzyme system. Generally these are synonymous with the term prosthetic group.

Coenzyme Q_{10} has been identified as the heart-healthy antioxidant. These are fat-soluble antioxidants. Q_{10} regenerates Vitamin E in the antioxidant network. For almost 25 years this was used successfully to treat and prevent heart disease in Japan and also in USA.

C_OQ_{10} rejuvenates brain cells, is currently used as a treatment for advanced breast cancer.

- Might prove effective in Alzheimer's and Parkinson's disease.
- Synthesized by the human body and is also found in seafood and organ meats.

C_OQ_{10} is gradually becoming popular as an antioxidant. It is not yet as well known as Vitamins E and C but there is a lot of potential for future use. It is present in all cell membranes. C_OQ_{10} is found in reasonable quantities in mitochondria, an energy producer in virtually every cell especially in the heart, brain, kidneys and liver i.e. in the most important parts of the body. Levels of Q_{10} decline with age. Even though this coenzyme is found in food such as salmon, liver and other organ meats, it is virtually impossible to get from diet alone and has to be supplemented. [2]

2.4.4 Vitamins as Antioxidants

The word vitamin might have originated from the words, 'vital amine' from an early belief that vitamins might have been all amines. It was soon dispelled and it was gradually established that vitamins have altogether different chemical, origins. It was found out that vitamins are essential to health and must be present in the diet and the body cannot make them. The term vitamin applies to any compound or a closely related group of compounds. Following are the salient features of a vitamin:

- It is organic rather than inorganic
- It's absence causes a specific vitamin deficiency disease
- It's presence is essential to normal growth and health.
- It is present in foods in small concentrations and it is not a carbohydrate, a saponifiable lipid, an amino acid or a protein.

Vitamins have been classified as fat-soluble or water-soluble. Fat-soluble vitamins are largely hydrocarbon like and water-soluble vitamins are polar or ionic.

Vitamins A, D, E and K are fat-soluble. These vitamins occur n the fat fraction of living systems. Water-soluble vitamins are Vitamin C, choline, Thiamine, Riloflavin, Nicotimide, Folic acid, Vitamin B_6, Vitamin B_{12}, Pantothenic acid and Biotin. These vitamins have been recognized and approved by the Food and Nutrition Board. It is also accepted that they are among the safest substances known. [7]

It was soon established that some of the vitamins act as antioxidants. Vitamins E, C and A have wide applications as antioxidants. These are examined here in some details.

2.4.4.1 Vitamin E

CAS No. [1406-18-4]

It is a general term referring to two groups of closely related 6-chromanol derivatives known as tocopherols and tocotrienols, each comprised of four homologs (α, β, γ and δ) that differ in the position and number of methyl substituents on the chromanol ring. All eight compounds are found in natural sources. Dietary sources include vegetable oils and green leafy vegetables. [4]

Most of the common tocopherols are found in Cereal Seeds, such as wheat, barley, rice but a rich source of tocotrienols is rubber latex. [8]

Vitamin E group of these tocopherols is typically represented as:

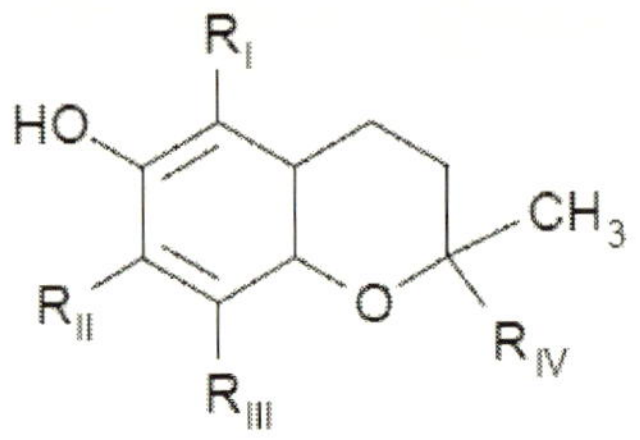

Tocopherols $R_{IV} = C_{16}H_{33}$

Tocotrienols $R_{IV} = C_{16}H_{27}$

$\alpha : R_I = R_{II} = R_{III} = CH_3$

$\beta : R_I = R_{III} = CH_3 R_{II} = H$

$\gamma : R_I = H \quad R_{II} = R_{III} = CH_3$

$\delta : R_I = R_{II} = H \quad R_{III} = CH_3$

Naturally existing tocopherols and tocotrienols exist in the unesterified forms and are distributed widely in nuts, seeds, fruits, vegetables and grasses. Major sources of Vitamin E in western diets are the cereal seed oils and the margarine and other products made from them. The indigenous concentration of Vitamin E in fatty foods are much higher than those of Vitamin A and Vitamin D. Availability of total Vitamin E in mg/100 gms is shown in Table 2.2.

Table 2.2 Food Sources of Vitamin E

Sources	mg/100 gm
Wheat germ oil	166 – 300
Soybean and cotton seed oil	45
Mayonnaise	58

The most potent and important tocopherol is α-tocopherol having the following structured formula and properties:

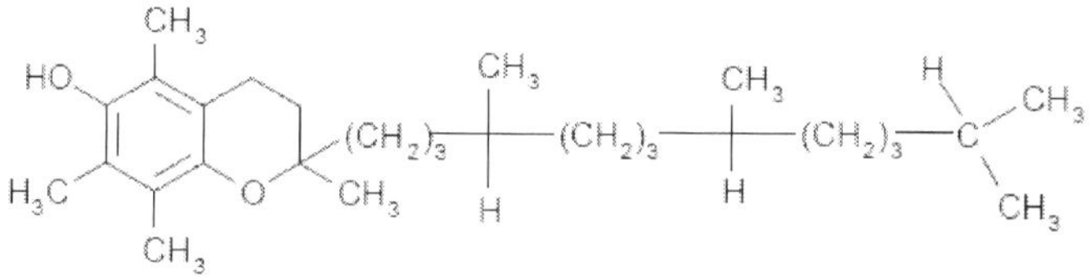

CAS No. [59-02-9]

mol. formula	$C_{29}H_{50}O_2$
mol. wt.	430.72
mp	45.6° C
bp	210 – 220° C / 0.1 mm Hg
d_4^{20}	0.950
η_D^{20}	0.5050

A clear yellow, viscous oil

Most of α-tocopherol and total Vitamin E though available in some quantity from natural sources, is grossly in adequate and is synthetically made. This is discussed in details in a separate chapter.

α-tocopherol is a natural biological antioxidant present in many vegetable oils is shown vide Table 2.3 [3,9].

Table 2.3 α-tocopherol in Vegetable Oils

Source (Vegetable oils)	**mg/kg**
Wheat germ oil	560 – 1200
Soybean oil	306 – 120
Sunflower oil	350 – 700
Walnut	560
Peanut	80 – 330
Coconut	5 – 10
Palm oil	180 – 260
Olive	1 – 240
Cotton Seed	40 – 560

Approximate α-tocopherol content of selected food products of animal origin is shown in Table 2.4.

Table 2.4 α-tocopherol in Selected Food Products [3, 9]

Source (animal products)	**α-tocopherol in mg/kg**
Beef	6
Chicken	4
Pork	5
Lard	12
Cod (liver oil)	12
Shrimp	7
Egg	5 – 11
Butter	10 – 33
Milk	0.2 – 1.1

Vitamin E in general and α-tocopherol in particular is gaining popularity in many fields. In fact it is the fastest growing vitamin.

Global demand of Vitamin E is projected at about 5000 tpa natural, and 15000 tpa synthetic, by the end of 2007.

2.4.4.2 Vitamin C (Ascorbic Acid)

CAS No. [50-81-7]

mol. formula	$C_6H_8O_6$	mp	190 –192°C
mol. wt.	176.3	d_4^{20}	1.65

Water soluble, strongly reducing, optically active (chiral) white crystalline substance is popularly known as Ascorbic acid.

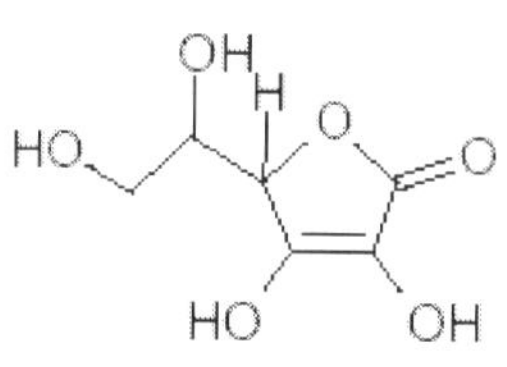

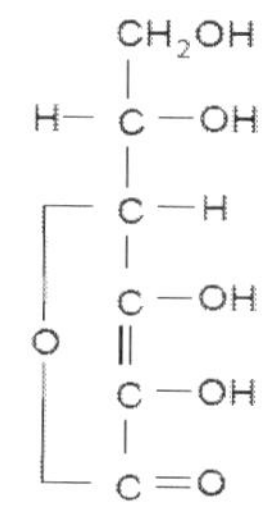

Vitamin C or L-Ascorbic acid

Insoluble in ether, chloroform, benzene, petroleum ether, oils, fats. Soluble in water and widely distributed in nature and a very popular physiological antioxidant, Dietary sources include citrous fruits, Potatoes, peppers, broccoli, cabbage and rose hips.

Inadequate intake results in deficiency syndromes such as Scurvy. [4]

Isomer of L-Ascorbic acid is D-Isoascorbic acid (Vitamin C).

Erythorbic acid or Isoascorbic acid, CAS No. [89-65-6], chemical formula $C_6H_8O_6$, mol. wt. 176.12, forms shiny granular crystals from water, alcohol, pyridine, moderately soluble in glycerol. It is used as an antioxidant and antimicrobial agent for foods. Chemical structure of Erythornbic acid is

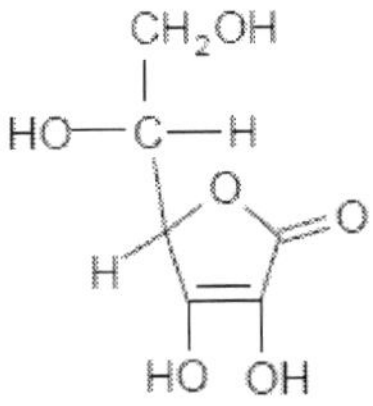

i) Erythorbic acid does not occur naturally

ii) Its biological activity is only 5% of that of ascorbic acid

iii) Due to its close chemical relationship to ascorbic acid it interferes with a number of biological functions of ascorbic acid

iv) In most of the European countries it is not permitted as a food additive

v) Various physical properties are different as shown in Table 2.5.

Table 2.5 Comparison of Physical Properties of Ascorbic Acid and its Isomer

Properties	Ascorbic acid	Erythorbic acid
Mol. wt.	176.13	176.13
Melting pt. range	190 – 192°C	164 – 169°C
Solubility gm/100 ml water Free acid	30	40
Optical rotation $[\alpha]_D^{20}$ (H_2O, C = 1)	+21°	-17°

Vitamin C is the most popular nutritional supplement in the market. One of the reasons is Vitamin C's ability to help the human body to develop an immunity for cough and cold. Also it helps to cure cough and cold which theory was postulated by Dr. Linus Pauling, the two-time Nobel Prize Winner (his first award was for chemistry). He also received his second Nobel award for peace. His book on Vitamin C and the common cold was very well received by one and all. [34]

Vitamin C acts as an antioxidant even for cancer. Dr. Pauling wrote about his experiences treating cancer patients in his last book, cancer and Vitamin C, along with another vitamin and a strong antioxidant Vitamin E, Vitamin C protects against the oxidation of lipoproteins. Vitamin C also plays a critical role in preserving the health of the cardiovascular system. Patients suffering from Diabetics need more Vitamin C. There is also growing evidence that Vitamin C may reduce the risk of developing cataracts. Vitamin C along with other antioxidants plays a role in protecting vision by combating cataracts. [2]

Blood pressure also seems to be affected by Vitamin C. A study at the Medical College of Georgia, in Augusta, found an association between high levels of Vitamin C in the blood and lower blood pressure. People with high levels of Vitamin C would have lower pressure than those with low intake of Vitamin C. And finally Vitamin C might restore fertility in some men. [11]

Vitamin C is abundantly available both in animals and plants. The Vitamin C content of some foods in mg/100 gm is shown in the following table: [12]

Table 2.5 Vitamin C contents in some foods (in mg/100 gm)

Source	Content	Source	Content
Fish	0 – 3	Asparagus	15 - 30
Liver, Kidney	10 – 40	Beans	10 - 30
Meat (beef, pork)	0 – 2	Broccoli	70 - 160
Milk (human)	3 – 6	Brussels sprouts	90 - 150
Milk (Cow)	1 – 2	Cabbage	30 - 60
Acerola	1300	Carrots	5 - 7
Apples	10 – 30	Cauliflower	60 - 80
Bananas	5 – 10	Oat rye, wheat	0
Black Currant	150 – 230	Parsley	170
Cherries	10	Potatoes	10 - 30
Guava	300	Peas	10 - 30
Hawthorn berries	160 – 800	Peppers	125 - 200
Lemons	50 – 80	Rice	0
Oranges	40 – 60	Spinach	50 - 90
Rose hips	1000	Tomatoes	10 - 30
Strawberries	40 – 90	Turnips	15 - 40

Amla is a fruit (fresh or dried) of the family Euphorbiacean and is known as Emblica Officiates. Amla fruit is a rich source of Vitamin C, 10 gms of fresh fruit contain 600 - 900 mg of Vitamin C. Small fruits contain more Vitamin C than large fruits. According to recent work fruits lose Vitamin C more when dried in the sun than in the shade used in curing of cough, bronchitis, Laemoptysis tuberculosis and is rejuvenating tonic because of Vitamin C content. [10]

Some salts of ascorbic acid such as calcium ascorbate [5743-27-1] and sodium ascorbate, [134-03-2] $C_6H_7NaO_6$, mw 198.11 are used as antimicrobial and, more importantly being relevant in this book, are used as antioxidant in food stuffs.

The most well known industrial uses of Vitamin C are in food industry and also in pharmaceuticals including multivitamin preparation, tablets, syrups elixirs and effervescent tablets. Antioxidant properties are exhibited in retardation of oxidative rancidity in fats and oils, increased clarity of wine and beer, synergistic activity in applications with other antioxidants (such as tocopherols). Further uses in the polymer, photographic and cosmetic industries are less important. [12]

Global demand of Vitamin C (ascorbic acid) from natural sources is projected at about 10000 tpa by the end of 2006. However, demand is steadily increasing as more and more new uses as an antioxidant are found. Vitamin C is also synthetically made mostly as a supplement. Now a days the trend is to establish large volume synthetic ascorbic acid (Vitamin C) plants since Vitamin C is not only used as an antioxidant or for pharmaceutical formulations but also for making some downstream products. Correct demand projection is therefore very difficult, published data and information from many countries is not exactly accurate and not reliable. This is discussed in Chapter 3.

Global manufacturer's include Hoffman La Roche, Dalry (UK) Takeda Ind. Chem. Osaka (Japan), Merck Darmstaat, BASF (Germany) etc. This will be again discussed in Chapter 3 under the caption Vitamin C which is made synthetically.

2.4.4.3 Vitamin A

CAS No. [11103-57-4]

The term Vitamin A refers to a group of monocyclic diterpenes with similar biological activity. The term Vitamin A is frequently linked to Vitamin A alcohol, retinol, retinoids including retinoic acid. Structurally these are as follow:

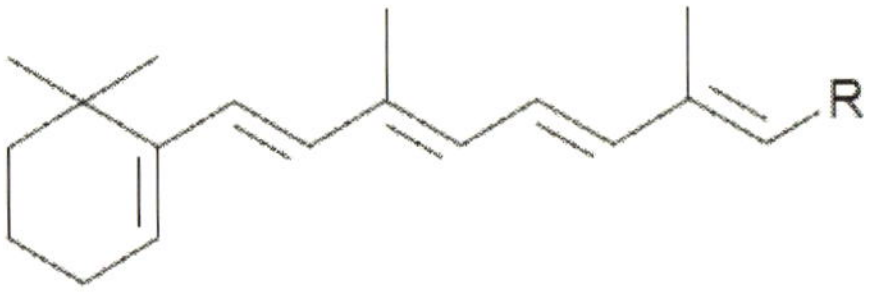

$R = CH_2OH$ retinol [68-26-8]

$R = CHO$ retinal [116-31-4]

$R = CO_2H$ retinoic acid [302-79-4]

Vitamin A activity is given by several vitamins. In the body the active forms retinol, retinal and retinoic acid – all have five alkaline groups, β-caroteins and to some extent, other carotenoids are precursors to Vitamin A (provitamins) and are described separately. A provitamin is a compound that the body can change into an active Vitamin A. Vitamin A is found only in animal tissue, whereas the Crotenoids i.e., provitamin A are found in the plant world. The latter are oxidatively degraded to Vitamin A compounds in the animal organism. [12]

Vitamin A is an example of nutrition therapy with a common affliction: Wounds. If one has a cut somewhere in the body perhaps with a kitchen knife or has burnt his fingers, studies suggest that Vitamin A might be able to speed your recovery. It is said that unlike many of the other Vitamin therapies, the healing effect of Vitamin A on wounds may be available without a need to turn to supplements. Vitamin A is particularly effective on patients who are taking steroid drugs. Steroids are often prescribed to control inflammation, but they also make skin slower to heal. Steroids or no steroids, eating a diet rich in Vitamin A such as carrots, squash, yams, cantaloupe and leafy green vegetables might be of great help.

Vitamin A also has a great role in the area of cancer control. A few studies suggest that Vitamin A may heal precancerous sores in the mouth called lenkoplakias. Vitamin A also is effective against measles, a childhood infection. [11]

Vitamin A deficiency causes numerous symptoms such as disturbances of vision (night blindness). Deficiency Syndromes can arise due to a deficit of carotene caused by insufficient protein supply.

German recommendations for the Vitamin A supply of working and domestic animals are shown vide Table 2.6.

Table 2.6 Recommended dietary allowances (RDA) for working and domestic animals

Animals	**Vitamin A equivalent to I.U., retinol per kg of dry feed per animal**
Piglets	10,000 – 16,000
Porkers	5,000 – 10,000
Milk cows	80,000 – 1,20,000
Horses	30,000 – 60,000
Chicks	8,000 – 12,000
Layers, turkeys	8,000 – 12, 000
Dogs	8,000 – 12,000
Cats	15,000 – 25,000
Fish	8,000 – 12,000
Beef cows	40,000 – 70,000

For humans, daily doses requirement vary with age. [12] Vitamin A requirement of humans is given as follows:

1 mg Retinol equivalent = 6 mg Vitamin A

1 RE = 3333 I.U.

mg retinol equivalent is the official unit of Vitamin A activity.

Table 2.7 Vitamin A Requirement of Human in RE (retinol equivalent)

Age	Retinol mg RE		
	m	m/f	f
10 – 13 years	0.9		0.9
13 – 15 years	1.1		1.0
15 – 19 years	1.1		0.9
19 – 25 years	1.0		0.8
25 – 51 years	1.0		0.8
51 – 65 years	1.0		0.8
> 65 years	1.0		0.8

Retinol is sold almost in the form of more stable esters mainly as retinyl acetate, also as retinyl Palmitate and Propionate. About 80% of production is used in animal feed, the remainder is used in the food and Pharmaceutical Sectors. [12]

Vitamin A acetate [11098-51-4] is the commercially significant form of the vitamin and is mainly produced by Hoffmann-La Roche, BASF, and Rhone-Poulenc. All of these processes have β-ionone as their key intermediate.

Vitamin A palmitate [79-81-2], a commercially important form of the vitamin is produced from Vitamin A acetate via trans-esterification reaction with methyl palmitate. Enzymatic preparation of the palmitate from the acetate has also been described.

Sulfone chemistry plays an integral part of the syntheses of both β-carotene and Vitamin A workers at Kuraray. Variation of Vitamin A content in food can also result from food processing and in particular, oxidation processes. [13]

Cod liver oil (mostly made in UK) and Shark liver oil (mostly made in coastal area cities in the world including India) are rich in

Vitamin A. According to Indian Pharmacopoeia 1 gm of Shark liver oil contains 6000 I.U. of Vitamin A. Shark liver oil should be stored in a airtight, completely filled closed containers protected from light.

Lemon oil, in particular, contains 1 to 2 percent volatile oil which contains 70 to 80 percent citral. Citral is usually used in the synthesis of β-ionone, which is used for synthesis of Vitamin A. This has resulted in importance of Lemon grass oil. It may be added that India produces good amounts of β-ionone synthesized from citral and is a leading exporter of both citral and β-ionone. [10]

Global demand of Vitamin A (excluding Carotenoids) has been projected at > 2000 tpa.

2.4.5 Phytochemicals

Phytochemicals are natural compounds found in fruits and vegetables. Each type of fruit and vegetable may contain hundreds of phytochemicals. An orange alone may contain 170 or more different phytochemicals.

Phytochemicals, at least many of them, have antioxidant properties. This means that they protect against "free radicals" which, as stated earlier, can damage healthy cells.

They appear to protect against arteriosclerosis – the build up of fatty plaques on the artery walls that can be risky and may cause heart attack and stroke – and against certain types of cancers. Allium contained in onions and garlic, has been under probe for its potential to reduce cholesterol levels. Ellegic acid, found in the berries may curb healthy cells from cancer. It may also protect the brain as it ages.

Carotenoids and Flavonoids being plant chemicals are also classified under phytochemicals. Flavonoids, in particular, also belong to polyphenols, also a part of the phytochemicals. Most of them have antioxidant properties.

Many of these photochemical offer unique benefits. Cruciferous vegetables such as broccoli have special chemicals that stimulate the production of anticancer compounds in the body. Soybeans are rich in isoflavones, hormone like compounds that fight cancer. Citrous fruits also contain cancer-fighting oil in their skin. Many of these phytochemicals are in the pigments of plants. Dark-green leafy vegetables and yellow and orange vegetables have different phytochemicals.

Most of the phytochemicals apart from food products include alkaloids, glucosides, terpene hydrocarbons, and glycerides (fats and oils). Some of these products have end uses in industry (pharmaceuticals, food, perfume, flavoring), there are also miscible plant products such as drugs, and pigments and plant hormones or growth regulators. Many of them have shown antioxidant properties. [1, 2] Phytochemicals are found in a variety of sources. Some phytochemicals which are extensively being studied for their antioxidant activities and ability to reduce disease risk are listed vide Table 2.8. [5]

Table 2.8 Phytochemicals and their Food Sources

Phytochemical	Food Source
Allyl Sulfides	Onions, garlic, leaves
Carotenoids (lycopene, lutein, zeax-anthin)	Tomatoes, Carrots, Watermelon, Spinach etc.
Carcumin	Turmeric
Flavonoids (anthocyanins, resvera-trol, catechins)	Grapes, Blueberries, Strawberries, Cherries, Applies, Cranberries, Raspberries, Blackberries etc.
Glutathione	Green leafy vegetables
Indoles	Broccoli, Cauliflower, Cabbage, Brussel Sprouts etc.
Isoflavones	Legumes (Peas, Soybean)
Lignans	Seeds (flax seeds, Sunflower seeds)
Monoterpenes	Citrons fruit peels, cherries, nuts
Phytic acid	Whole grains, legumes
Phenols, polyphenols, phenolic compounds such as ellagic acid, tannins, ferulic acid etc.	Grapes, blue berries, strawberries, cherries, tea
Saponins	Beans, legumes

These antioxidant-rich phytochemicals particularly fruits and vegetables should be a must for human diets. Of course in addition to food items mentioned earlier getting regular exercise and abstaining from tobacco and excessive alcohol drinking, should prove to be critical to a healthy lifestyle which would retard the process of aging and premature deaths.

Projecting global demand by the end of 2006 including Carotenoids, Flaronoids and others should not be an easy task. However, collectively it is to the tune of 4000 – 5000 tpa.

2.4.6 Flavonoids

Flavonoids are a group of aromatic, oxygen containing, heterocyclic pigments widely distributed among higher plants. They constitute most of the yellow, red and blue colors in flowers and fruits.

Carotenoids are the exceptions. The flavonoids include the following sub-groups: [1]

1. Catachins
2. Lenco anthocyanidins and flavones
3. Flavanins, flavones and anthocyanins
4. Flavonols

Flavonoid molecules are polyphenols. A large number have been identified, though with varying metabolic properties. They were controversial for some time though now they have been accepted in some countries as playing an auxiliary role in the redox function of Vitamin C. They have now been accepted as antioxidants. Following three conclusions have now been arrived at

- the speculation that flavonoids are inactive has been rejected
- Studies carried out on specific studies have established the important of flavonoids, at least some of them
- Though earlier considered as controversial, the unquestionable evidence of very promising biological and chemical effects from favonoids deserves attention and further research work. Their antioxidant properties to scavenge the appropriate free radicals are known. [3]

Bioflavonoids, which are included in foods as flavorings, have been recognised as effective in decreasing the erythocite aggregation and sedimentation rate of blood in vitro and are useful in dietary control for problems related to coagulation of platelets.

These high molecular weight polyphenolic antioxidants possess anticarcinogenic, antiproliferation, antimetastatic and prodiffentiation activities. An increased intake of these antioxidants can contribute to minimizing the problem of oxidation of dietary cholesterol, which is the cause of cardiac arrest.

So far 4000 flavonoids have been identified though 50 of them have been studied extensively in various laboratories in the world. These flavonoids are found in cranberries, onions, broccoli, celery, soybeans, tomatoes, eggplant, cherries, applies, red wine and grape juice.

Tea and red wine are rich sources of antioxidants of the flavonoid group paradoxically numerous studies have documented that people who drink red wine, which is particularly high in flavonoids, have much lower rates of heart disease than those who don't drink wine. Similarly tea, particularly green tea, (very popular in China and Japan) contains flavonoid-antioxidants and is strongly recommended for good healths and slows down the process of aging.

In a recent publication of the Indian Daily, The Times of India edition of September 14, 2006, Mumbai, a write-up regarding green tea emphasized the usefulness of green tea. It was brought to light that adults who drank at least five cups of the brew daily had a lower risk of Cardio-Vascular deaths (excluding death from cancer) then those who drank less than a cup a day. These findings were based on research at Tohuku University in Japan.

Experts believe natural compounds called polyphenols – particularly one substance called epigallocatechin-3-gallate, provide the health benefits. These substances have strong antioxidant properties that may help prevent heart disease.

Two most important flavonoids, namely, Gingko biloba extract and Pycnogenol (Pine bark extract) were soon found to be very powerful antioxidants. A combination of Ginkgo and Pycnogenol proved to be more powerful than their individual components because of Synergistic activities. These two groups are accordingly discussed in details. It was also discovered that flavonoids interact with other well-known antioxidants such as Vitamin C and Vitamin E. Flavonoids regenerate Vitamin C by boosting its activities. Vitamin C in its turn increases levels of Vitamin E. In other words both water soluble (Vitamin C) and fat soluble (Vitamin E) antioxidants are rejuvenated by flavonoids.

2.4.6.1 Gingko biloba

The Gingko is of stone age (some 5000 years ago). Once very common in North America and Europe it was destroyed during the ice age everywhere on earth except in China, where it still lives and thrives. It was brought to America in 1784 and is now one of the most common trees in USA. It is also cultivated in Japan.

Gingko is a strong, hardy tree that can grow as tall as 122 ft and 4 ft in diameter. Some gingkos have lived to be 1000 years old. Ancient healers took note of the strength and vigor of the gingko. The leaves and fruit of the gingko have been used for medicinal treatment in China since 2800 B.C. primarily for treatment of damage in brain cells and other circulatory disorders.

Ginkgolides belong to a family of terpenes isolated from the root bark and leaves of Ginkgo biloba L. Ginkgoaceae leaves Ginkgolide A, CAS No. [15291-75-5], mw 408.40 and mol. formula $C_{20}H_{24}O_9$, Ginkgolide B, CAS No. [15291-77-7] mol. formula $C_{20}H_{24}O_{10}$, mw 424.40, and Ginkgolide C, CAS No. [15291-76-6], $C_{20}H_{24}O_{11}$, mw 456.40, have been isolated. Ginkgolide B has been used in the treatment of Severe Sepsis [3, 4].

Ginkgo biloba extract along with its derivatives have been identified as extremely useful antioxidant. Global demand has been projected at about 5000 tpa by end 2007.

Gingko is a Maidenhair tree, also known as Kew tree, is the only living member of the Ginkgoacese. Valued as a street and park tree, the ripe fruit gives off a foul odor. The seeds and leaves have been used in traditional medicine for asthma and vascular disease. A wide variety of compounds have been identified in the fruit, leaves and bark including Ginkgolides, bilobalide, bioflavones, flavonal glycosides. Fruit pulp contains toxic principles: Ginkgolic acid and Ginkgol.

Ginkgolides have been extracted from root bark. (Extraction Gingko biloba EGI 761)

- A defined extract obtained from the leaves, stated to be noortropic used, in vascular insufficiency.

2.4.6.2 Pycnogenol

Though Pycnogenol, an extract from the bark of French maritime pine tree has been used in the past for many years for traditional healers. However, recently as a result of modern science and technology this was rediscovered and reestablished as a powerful antioxidant. From ancient times pine bark has been used as a source of food and medicine. The pounded inner pulp of the pine tree bark was used for inflammatory disorders and wounds, sores and ulcers. It's a very potent antioxidant containing proanthocyanides.

Pycnogenol is also used for boosting Vitamin C activity. It was proved by Scientists of Packer laboratory, Berkeley that Pycnogenol can boost the body's production of other antioxidants, such as Vitamin E and glutathione further increasing cell defenses against free radicals attack. [2] Pycnogenol (Procyanidin extract of the bark of Pinus maritima) is probably one of the most studied phenolic tree extracts. It has been shown to Scavenge free radicals, including hydroxyl and super oxide anions. It is stated also to have beneficial effects in preventing atherosclerosis and other age-related diseases. [23]

Pycnogenol may protect against Cardio-Vascular disease, strengthens the immune function, fights against chromic fatigue syndrome and is believed to play an active role in slowing down aging. Also it helps tremendously as a potential treatment for attention-deficit hyperactivity disorder. As in the case of another flavonoid Ginkgo biloba, Pycnogenol is used in treatment of memory related problems in a big way.

2.4.7 Carotenoids

Carotenoids are red, yellow and orange pigments which are widely distributed in nature. Although specific carotenoids have been identified in photosynthetic centers in plants, bird feathers, crustaceans and marigold Petals, they are especially abundant in yellow-orange fruits and vegetables and dark-green leafy vegetables. So far 700 naturally occurring carotenoids have been identified, however, only 50 are stated to be absorbed and metabolized by the human body. Hydrocarbon carotenoids are classified as carotenes and those containing oxygen also are known as Xanthophylls. Fig.2.2 shows the prominent carotenoids.

More than 500 different molecules have been identified as carotenoids. A major part of natural colors constituents the carotenoids or the yellow and orange pigments (natural colors). Cartonoids are mainly of two types: (1) Carotenes, (2) Xanthophylls. Xanthophylls has originated from the Greek words xantho meaning 'yellow' and 'phylls' indicates from leaves. It has a wide range of uses in the fields of medicine, pharmaceuticals, nutracenticals and food industry.

Neutracenticals encompass phenolic or polyphenolic compounds and are functionally important as natural antioxidants as they hinder or retard the process of formation and propagation of free radicals.

Carotenoids are essential to photosynthetic processes in terrestrial and marine plants. What is most important and of Prime interest to the readers is that carotenoids exhibit high antioxidant activity.

Carotenoid molecules corresponded to isoprenoid polyenes ($C_{40}H_{56}$). They are constituted by eight Cis-isoprene units and have 22-c Central Skeleton, and an additional 9 carbon atoms in each molecule.

The presence of double conjugated bonds in the structure of carotenoids is responsible for their pigmentation, antioxidant activity and high susceptibility to oxidation and geometric isomerization in the presence of oxygen, light, acids and /or organic solvents. In plant material, however, carotenoids are commonly associated to lipids. They are thus stable even at high temperature, are soluble in non-polar solvents such as petroleum ether, normal hexane etc. Some of the Carotenoids are discussed here.

2.4.7.1 Lycopene

Long deep red needles from a mixture of carbon disulfide and ethanol, also from a mixture of methylene chloride and methanol.

Also known as ψ, ψ-Carotene (all trans).

A long chain hydrocarbon with 13 double bonds of which 11 are conjugated combustible.

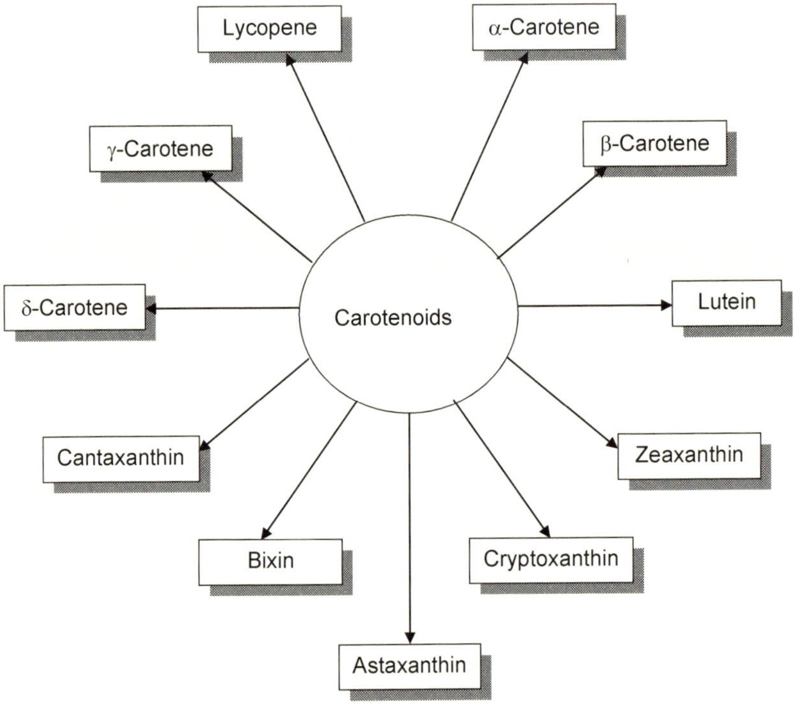

Fig. 2.2 Prominent Carotenoids

CAS No. [502-65-8]

mol. formula	$C_{40}H_{56}$
mol. wt.	536.87
mp	172 – 175° C

Also soluble in Acetone, Petroleum ether. Insoluble in water, Ethanol, methanol.

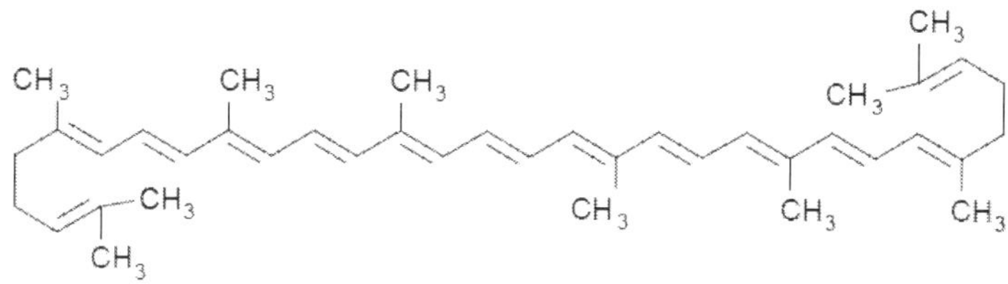

Lycopene, the red pigment of the tomato, is a C_{40}-carotenoid made up of eight isoprene units.

Lycopene is a natural pigment synthesized exclusively by plants and microorganisms. One of the functions of lycopene and related carotenoid species is to absorb light during photosynthesis thereby protecting plants against photosensitization. Sometimes the green chlorophyllic pigments mask the brilliant colors of lycopene. The chlorophyll content decrease as the plants mature, leaving the lycopene and other carotenoids responsible for the bright colors of most fruits such as pine apple, tomato, paprika, rosehip and many flowers.

Lycopene is recognized to be the most efficient singlet oxygen quencher among biological carotenoids. Antioxygen functions are associated with lowering DNA damage and reducing biological oxidative damage of proteins, lipids and other cell components in vitro. There is found protective effects of lycopenes towards oxidative stress-mediated damage of the human skin after irradiation with UV light. Lycopene has preventive effects on atherosclerosis by protecting plasma lipids from oxidation. Lower blood lycopene levels are also associated with increased risk of coronary heart disease.

It is suggested that daily consumption of tomato products providing at least 40 mg of lycopene is enough to substantially reduce LDL oxidation. [14]

Recently interest in the consumption of lycopene rich foods as a means of reducing the risk of cancer has prompted researchers to investigate the level of lycopene in foods frequently consumed by the people, although it has no provitamin A activity. Lycopene in able to function as an antioxidant and exhibits a physical quenching rate constant with singlet oxygen in vitro.

Cooked tomatoes are a better source of lycopene than raw ones. It was found that lycopene in tomato paste is better utilized by the body than whole tomatoes. It was also stated that since lycopene is fat soluble, it is best absorbed by the body when eaten in mealtime with a bit of oil. Lycopene is present in many fruits such as watermelon, tomato, paprika, rosehip and also in many flowers. Tomato is the richest source of lycopene whereas watermelon is the second richest source. Lycopene in tomato varies from 5-40 mg per 100 gm of tomatoes. [14]

Watermelons and pink grapefruit are other sources of lycopene.

Total lycopene content of some familiar selected fruits is shown by the following table: [15]

Table 2.9 Total Lycopene content in Selected Fruits

Source	Lycopene μg/100 gms
Apple	60 – 308
Banana	16 – 247
Guava	769 – 1816
Jack fruit	37 – 111
Mango	49 – 724
Orange	0
Papaya	4305 – 7564
Pineapples	265 – 605
Tangerine	222 – 670
Red Watermelon	8731 – 13523
Yellow watermelon	0 –109
Star fruit	16 - 94

Lycopene has assumed tremendous importances both to industry and research and academic institutions.

2.4.7.2 α-Carotene

CAS No. [7488-99-5]

mol. formula $C_{40}H_{56}$
mol. wt. 536.87

Carotenes have four important isomers, α, β, γ and δ apart from Lycopene.
As widely distributed as its β-isomer but in smaller amounts.

Best source for both α- and β-isomers and small quantities of γ-isomer are carrots, palm oil and green leaves of various species.
- As a provitamin A it is half as active as β-carotene. Found in the mother liquors after crystallization of β-carotene. It is isolated by chromatography.

Structure is as follows:

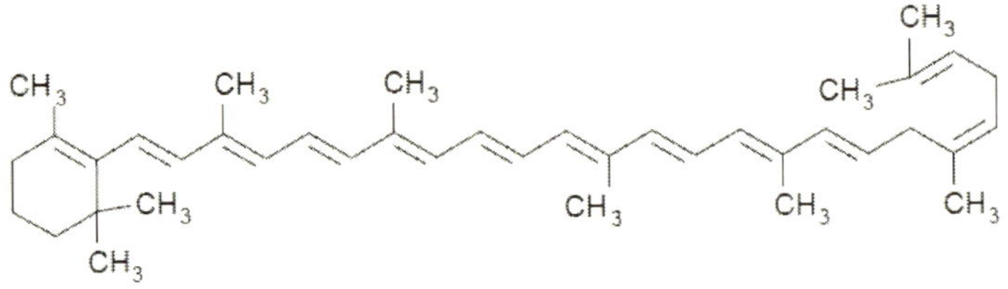

Deep purple prisms, polyhedra from petroleum ether or from benzene and methanol. m.p. 187.50°C. Freely soluble in CS_2, chloroform, soluble in ether, benzene. Absorbs oxygen from air giving rise to inactive colorless oxidation products. The oxidation in presence of light is autocatalytic.

Uses:

- Vitamin A precursor (Therapeutic)
- Vitamin A precursor for all species except cats (Veterinary use)

β-Carotene being the most dominant carotene in some fruits and vegetables, all other carotenes is much less significant. However, α-carotene is most important next to β-carotene in the carotene family.

2.4.7.3 β-Carotene

CAS No. [7235-40-7]

β-Carotene is most important both as a provitamin A and as an antioxidant in the group of carotenes. So much so people mean β-carotene whenever they talk of carotenes. In plants and animal Kingdom this is widely distributed. In plants it occurs almost always with chlorophyll. Though many green leafy vegetables, roots and fruits have good content of carotenes in general and β-carotene in particular, the best source of β-carotene and to some extent α-carotene is carrots. Carrots occupy a pride of place among fruit and vegetables so far as nutrition is concerned. No other vegetables or fruits contain as much carotene, which the body is able to convent to Provitamin A. β-carotene is also an excellent antioxidant. The high level of β-carotene is responsible for imparting the distinctive orange color to the vegetable i.e., carrot. The beta-carotene in carrots is a natural protector for common skin cancer and melamine. Certain eye diseases that Vitamin A and β-carotene have been known to cure are:

- Night blindness
- Conjunctivitis / Blepharitis
- Retinopathy
- Cataracts particularly true of β-carotene and phytochemicals.
- Helps in preventing cancer by annual cell reproduction.

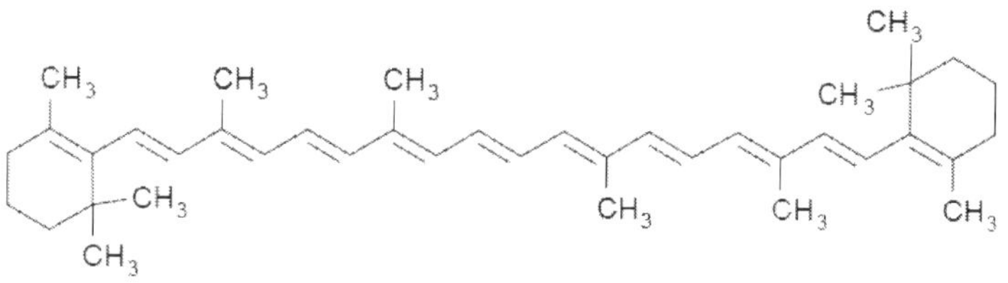

Molecular formula $C_{40}H_{56}$, mol. wt. 536.87

β-Carotene the yellow pigment of the carrot is also an isomer of lycopene, which has been discussed earlier.

Carotenoids, including β-carotene, is very sensitive to photochemical air oxidation.

There is an increasing interest in the possibility of health benefit of antioxidants. The antioxidant activity of β-carotene has been associated with reduced risk of lung, colon and various types of cancers such as stomach, esophagus and possibly cancer of the mouth, bladder, prostate, ovary, breast and the cervix as well as reducing the risk of Coronary heart diseases. β-Carotene has a very high antioxidant value and acts as free radical scavengers thus preventing various organs including heart disease β-carotene also helps by destroying the free radicals after they are formed. It has been implicated that β-carotene is a very effective antioxidant against the highly destructive hydroxyl (OH) radical also. [14]

Total carotene and β-carotene content of common Green Leafy Vegetables (GLV) have been shown in Table 2.10. [16]

Table 2.10 Total Carotenes and β-carotene content of Some Green Leafy Vegetables (GLV)

Food Stuff – Local / botanical name	Total Carotene mg/100 gm	β-carotene	
		mg/100 gm	% of total
A) GLV with high Vitamin A activity Drumstick (Moringa Oleifera)	42.1 ± 8.8	19.7 ± 5.5	6.3 ± 7.97
Agathi (Sesbanic grandiflora)	45.2 ± 5.54	15.4 ± 6.87	32.5 ± 11.5
Methi (Trigonella foenum graecum)	11.8 ± 3.70	9.2 ± 1.48	81.0 V 10.71
Amaranth (Amaranthus gangeticus)	21.2 ± 5.49	8.6 ± 2.78	39.7 ± 3.76
Curry leaf (Murraya Koenigil)	21.4 ± 3.44	7.1 ± 2.36	31.4 ± 5.99

Table 2.10 Total and β-carotene content of Some Green Leafy Vegetables (GLV) (contd.)

Food Stuff – Local/ botanical name	Total Carotene mg/100 gm	β-carotene	
		mg/100 gm	% of total
Gogu (Hibiscus Sabdariffa)	15.8 ± 3.58	5.8 ± 2.77	34.5 ± 11.1
Pudina (mint) (Menta arvensis)	19.1 ± 5.0	4.3 ± 2.0	21.0 ± 5.44
Chammakura (Cococasia antiquoram)	19.1 ± 4.24	5.5 ± 1.86	28.9 ± 6.82
Koyyalakura (Sueda monoica)	17.1 ± 0.70	4.8 ± 0.63	28.1 ± 2.96
Onion leaf (Allium cefa)	9.8 ± 0.39	4.9 ± 0.15	49.7 ± 1.54
Coriander (Coriandrum sativum)	15.2 ± 1.30	4.8 ± 0.16	31.6 ± 1.14

Apart from the vegetables mentioned above some other vegetables containing carotene and β-carotene are Palak (spinach), reddish, lettuce etc. β-carotene was the only provitamin A carotenoid detected in the above vegetables shown in the table. Carrot, sweet potato, pumpkin, tomato ripe and capsicum are other vegetables containing high carotene (and particularly β-carotene).

Among the fruits mango and Indian dates are relatively high in β-carotene content. In papaya β-carotene is only 1.0 mg/100 gm but it contains significant amount of β-cryptoxanthin (1.04 mg / 100 gm), which has 50-60% of the potency of β-carotene. This enhances both Vitamin A and antioxidant value of papaya.

Very small amounts of other carotenes such as γ-, and δ-, CAS No.S [472-83-5] and [472-92-4] respectively, are present in the fruits and vegetables mentioned above.

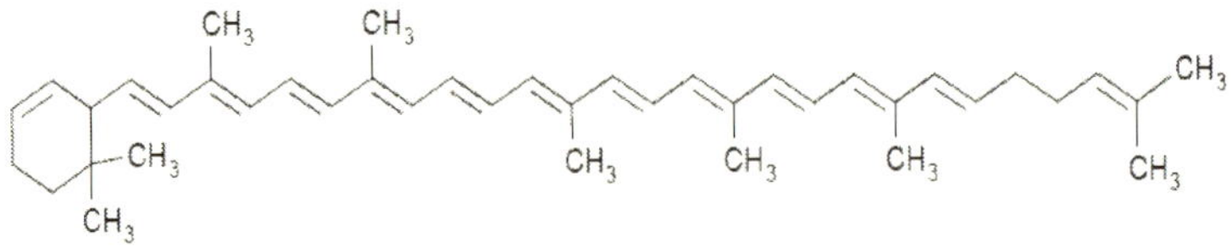

γ-Carotene

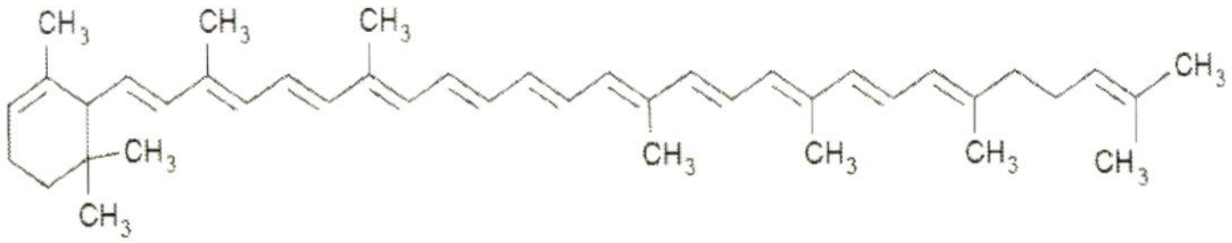

δ-Carotein

γ-Carotene and δ-carotenes are isomers of β-carotene and are associated with β-carotene containing fruits and vegetables in insignificant quantities and are unlikely to have commercial applications though they have been identified and isolated by chromatographic methods. All of them have same chemical formula ($C_{40}H_{56}$) and molecular wt. 536.87. Only more research work on these minor carotenes might focus more attention and establish the usefulness of these carotenes. [4] So far researchers have focused primarily on the important carotenoids, six of them, such as alpha-carotene, β-carotene, cryptoxanthin, lutein, lycopene and zeaxanthin, collectively, and not individually as Beta-carotens, they can do wonders. Fruits and vegetables rich in those carotenoids are known for their bright colors, ranging from brilliant yellow, red and orange to purple and dark green.

Though all of them have antioxidant properties only α-carotene, β-carotene, and cryptoxanthine have provitamin potency and are converted into Vitamin A in the human body. It is also now believed that the individual carotenoids may play a role in the body that is probably unrelated to Vitamin A.

Apart from many controversies about the carotenoids and their constituent units like carotenes, they do act as antioxidants in the body but functionally may not act similarly.

2.4.7.4 Lutein

Lutein is a carotenoid. It shows the same characteristics of an antioxidant as carotenoids.

Properties of Lutein are shown vide Table 2.11. [14]

Table 2.11 Properties of Lutein

Molecular formula	$C_{40}H_{52}O_2$
Mol. wt.	568.87
Melting Point	195°C
Appearance	Yellow paste
Solubility	Soluble in chloroform, h-Hexane, Benzene, CS_2, Methylene Chloride, Acetone, petroleum ether Insoluble in water, Ethanol Methanol
Sensitivity	Light, Oxygen, High temperature acids.

Structurally it is shown as follows: [17]

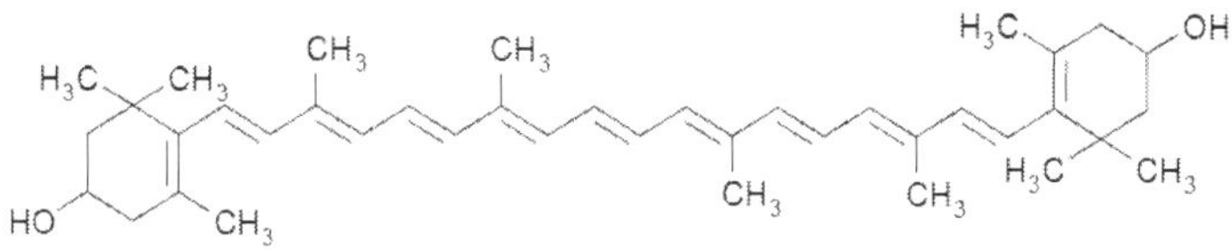

Marigold (Calendula Officinals) flowers contain flavonoids, carotenes and luteins. Marigold extract is used as a natural color in different food applications including as a blend with chicken feed to induce attractive color to the poultry eggs. Lutein and lutein esters have good potential as food colorants. Depending on the concentration of colorant the color varies from yellow to orange.

Applications of lutein are mainly in pharmaceutical, nutra-centical, cosmetic and food industries. Lutein, a member of the carotenoid family, acts as an antioxidant that protects healthy cells from free radical damage. Marigold containing lutein acts as an antioxidant for aiding in vision.

The two commonest causes of blindness in the elderly age group are cataract and macular degeneration. Macula is the spot on the retina that creates sharpest vision. When the macula is degenerated or goes for even so does the vision. The yellow color of the Macula is derived from Green vegetables, which are rich in carotenoids containing lutein. Clinical tests have shown the relationship between macular degeneration and carotenoid, particularly, lutein deficiency. Extraction of lutein and other carotenoids is discussed separately.

Lutein possesses chemo-preventive activity induces gap junction communication between cells and inhibits lipid peroxidation in vitro more effectively than β-carotein, α-carotene and lycopene. Apart from marigold, lutein also occurs in mango, broccoli, papaya, orange, kiwi, cabbage, spinach, lettuce, sweet potatoes etc. However, commercially it is obtained from the extraction of marigold. Lutein does not possess provitamin activity [18].

Lutein, a yellow pigment, has been isolated from the corpus luteum and found in body fats and egg yolks. It is also similar to or identical with Xanthophyll. Raw spinach contains high amounts of lutein, which as stated earlier plays an active role in preventing macular degeneration. A study undertaken by five prominent eye research centers in the United States showed that people who consumed the most lutein in their diet had a much lower incidence of macular degeneration than those who did not. Spinach was found to have a protective effect against the leading cause of blindness.

It can be safety concluded that future researchers will discover multiplicity of diversified uses of Lutein.

2.4.7.5 Carotenoids of the Xanthophyll Group

1) Zeaxanthin
2) Cryptoxanthin
3) Bixin

To start with, their properties are as follows:

	Zeaxanthin	Cryptoxanthin	Bixin
CAS No.	[144-68-3]	[472-70-8]	[6983-79-5]
molecular formula	$C_{40}H_{56}O_2$	$C_{40}H_{56}O_2$	$C_{25}H_{30}O_4$
mw	568.87	552.87	394.50
Solubility	practically insoluble in water, slightly soluble in petroleum ether, methanol. Soluble in CS_2, benzene, CCl_4 chloroform, pyridine, ethyl acetate glacial ac. acid+ hexane	Freely soluble in chloroform, benzene pyridine, CS_2. Less soluble in ligroin, petroleum ether, alcohol methanol	Considerably less soluble in organic solvents such as acetone, benzene, chloroform, acetic acid etc. than the labile form stable trans-form (also called isobixin)
Physical form	yellow rhombic plates with steel blue metallic luster	Red plates with metallic lustre from ether + methanol	Orange to purple plates from acetone

Chemical structures of the three compounds are shown below:

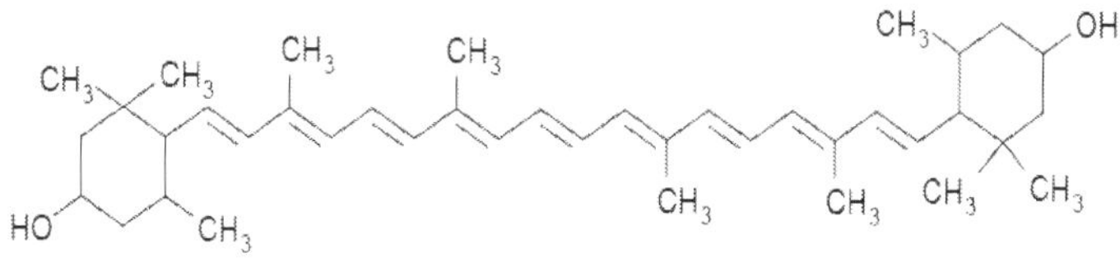

a) Zeaxanthin

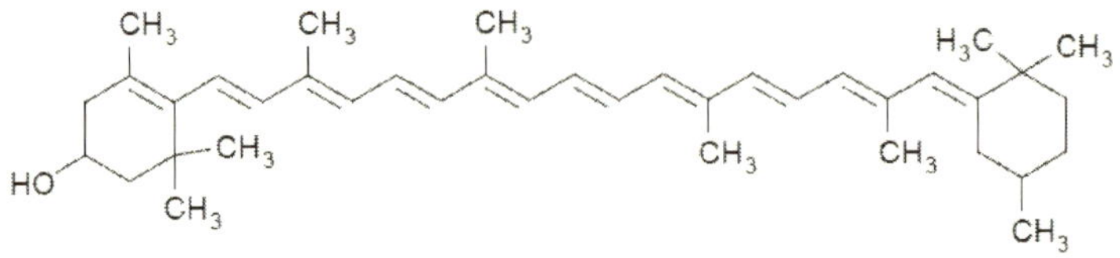

b) Cryptoxanthin

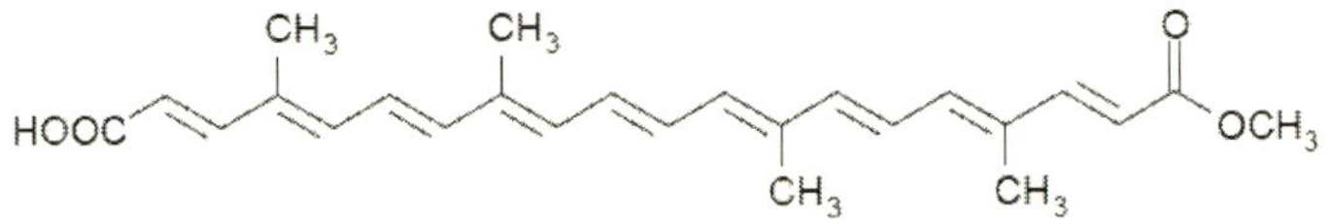

c) Bixin

Cryptoxanthin pigment with Vitamin A activity, Isolation from petals of certain flowers and berries of physilis species. Widely distributed it was also isolated from orange rind, egg yolk, butter, bovine blood serum.

As in the case of lutein zeaxanthin exists in the retina of the eye and confers oxidative protection against macular degeneration. Zeaxanthin is also prevalent in ovaries and adipocyte tissue. Dietary sources include peaches, squash, apricots. Oranges, papaya, prunes, pumpkin, kale, kiwi, lettuce, honeydew, melon and yellow corn. This xanthophyll does not possess provitamin A activity. [18]

Alcohol consumption has been shown to influence lipid peroxidation. Xeaxanthin has protective effects as an antioxidant against LDL oxidation.

Bixin is a carotenoil carboxylic acid from seeds of Bixa Orellana L. Bixacean. Bixin does not possess provitamin A active and is not widely known as an antioxidant. The cis-form occurring in the plant is unstable and changes to the stable trans form. Gets completely converted to the trans-form on standing in chloroform solution in the presence of iodine. In decomposes at 217°C. More applications and properties will probably be known to the future researchers.

2.4.8 Citric Acid

CAS No. [77-92-9]

Structure

$$HO - \overset{\displaystyle CH_2COOH}{\underset{\displaystyle CH_2COOH}{C}} - COOH$$

mol. formula $C_6H_8O_7$
mol. wt. 191.12
$HOOCCH_2C(OH)\ (COOH)\ CH_2COOH$
(2-hydroxy-1,2,3-propane tricarboxylic acid)

Widely distributed in plants and in animal tissues and fluids. IS also extracted from citrons fruits.

Citric acid is a weak chelating agent. However, it is very effective in retarding the oxidative deterioration of lipids in foods and is commonly added to vegetable oils after deodorisation. Citric acid is a good food preservative, presumably, because of its antioxidant qualities. [3]

Properties: Colorless, translucent crystals or powder, odorless strongly acidic taste. d_4^{20} 1.542 mp 153° C (anhydrous form). Decomposes before boiling, very soluble in water and alcohol, soluble in ether. Combustible made by fermentation of molasses, also from lemon, lime and pineapple used in preparation of flavoring extracts, preservative (soaps etc.) and an antioxidant. [1]

2.4.9 Uric Acid

CAS No. [69-93-2]	
Chemical formula	$C_5H_4N_4O_3$
mol. wt.	168.11
d_4^{20}	1.89
Chemical structure	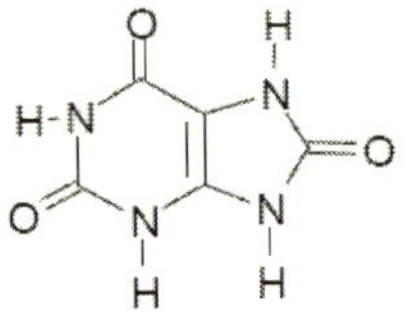

White, odorless, tasteless crystals, decomposes on heating without melting and with evolution of HCN.

It forms the chief end product of the DNA metabolism of birds and of scaly reptiles and is found in their increment.

Present in the urine of all carnivorous animals. May be synthetically prepared from urea. Partially soluble in water, soluble in glycerol in solutions of hydroxides and their carbonates, sodium acetate and sodium phosphate. Insoluble in alcohol and ether.

It is a minor antioxidant and does not have much use for commercial exploitation. In a recently published article on uric acid it was stated that uric acid is a breakdown product of high-protein food intake. In mammals it is stated to be a product of metabolism of purine compounds contained in non-vegetarian food.

Uric acid is supposed to increase by impaired kidney function caused primarily by diabetes, and also as a result of high intake of protein. Red meat, alcohol, lentils, custard apple chiku, spinach and peanuts are rich in purine compounds and can lead to high uric acid levels.

Problems caused by high uric acid levels create propensity to develop kidney stones, result in development of gout, a kind of arthritis caused by deposition of uric acid crystals in joints. Endothelial dysfunction that indisposes a person to develop hypertension, that in its turn can further inhibits kidney function. Drinking lots of water and regular exercise and cutting down on non-vegetarian foods can inhibit the ill factors associated with ureic acid. [19]

2.4.10 Guaiac Gum

CAS No. [9000-29-7]

Contains 70% α and β guaiaconic acids, 15% vanillin. It is a resin extracted from certain Mexican and West Indian trees, especially Guaiacum Santum and G. Officinale.

Properties: Brownish lumps, mp 85° C Insoluble in water, soluble in alcohol, ether, acetone, chloroform and caustic soda. It is used as a flavoring agent in foods, medicine (diagnostic aid for blood or hemoglobin). Also it has antioxidant properties and also used in food preservation. Has been used mainly to stabilize refined animal fats in combination with phosphoric acid derivatives as synergists. It cannot be regarded as a common foodstuff, but as an additive. It's use is limited to a few countries only. The antioxidant principles of guaiac gum are the α- and β-guaiaconic acids. [1, 3, 4]

3

SYNTHETIC ANTIOXIDANTS

3.1 CHEMICAL SYNTHESIS OF SOME ANTIOXIDANTS

In the preceding chapter natural antioxidants were discussed in depth. Most of the antioxidants derived from natural sources i.e., plants and animals are used for retardation of oxidative degradation of food products. These antioxidants are generally recommended keeping in mind safety (generally regarded as safe or GRAS) and also taking into consideration the after effects i.e., toxicity and development of carcinogenic properties after a defined and predetermined shelf life.

Collectively production of natural antioxidants from all possible sources would not exceed 40-50000 tpa. However, total demand of antioxidants is estimated to the tune of some 1,50,000 tpa. Hence it was considered necessary to supplement the natural antioxidants with synthetically made antioxidants. Besides, most of the natural antioxidants have been catering to the needs of food products. Synthetic antioxidants, however, would meet the requirements of all products including Food, Rubber, Polymers, elastomers and others wherever there is a possibility of degeneration of the product after a reasonable period. More than toxicity or carcinogeniety it is the life of the product, which is most important in many cases. These synthetically made antioxidants are mostly hindered phenols and polyphenols. They either counteract the negative effect of free radicals formed by auto-oxidation or are usually oxygen scavengers increasing the shelf life of the product, by removing oxygen from the system or by terminating the chain reaction process in a closed system.

Reaction of atmospheric oxygen with organic compounds leads to a number of degradative changes that reduces the lifetime of many products of interest to the chemical and food processing industry. This is especially true for petroleum products encompassing the polymers and other petrochemical derivatives, and also deterioration of lipids in foods. It was almost a Century ago that deterioration of rubber as a result of oxidative causes was noticed and demonstrated by the scientists who started studying the effect of oxidative degeneration of many more products and were working on the process of inhibition of such damages by suitable antioxidant devices.

Some of the more well known synthetic antioxidants including their chemistry are discussed in this chapter. The will cover many of the butylated phenolic and other types of antioxidants. In case of uses of such antioxidants for food products one has to keep in mind the toxicity and a long-term carcinogenicity that would affect badly the quality of stored and packaged food in general. Most of these antioxidants have been commercially exploited though many are yet to be commercially explored. Fig.3.1 shows the prominent synthetic antioxidants.

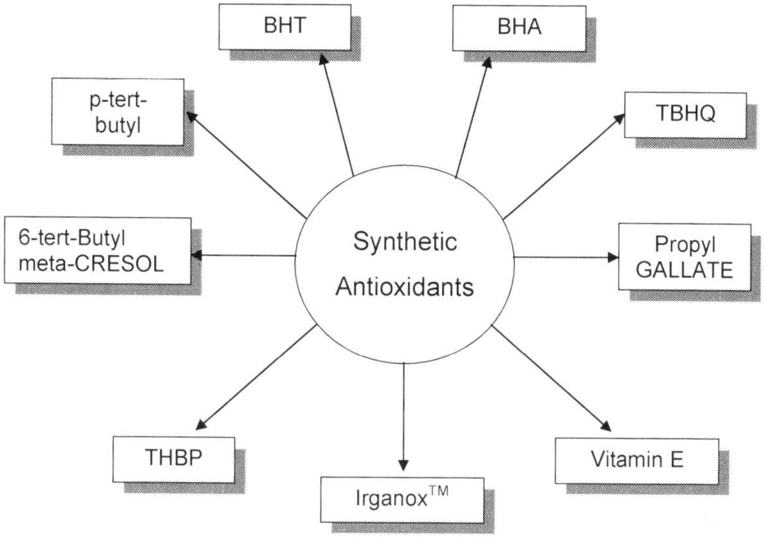

Fig. 3.1 Some Prominent Synthetic Antioxidants

3.2 para-tert-butyl phenol

CAS No. [98-54-4]

mol. formula	$C_{10}H_{14}O$
mol. wt.	150.22
Chemical structure	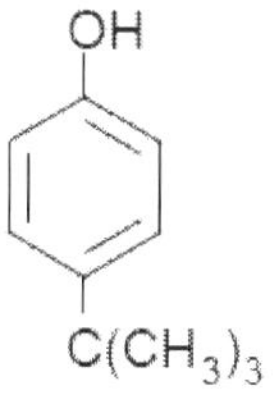
	Forms needles from water
mp	98° C
d_4^{114}	0.9081
bp	237° C

Manufacturing Process: Phenol with isobutylene using H_2SO_4 catalyst or heating phenol and isobutanol in the presence of zinc chloride catalyst. Is also made by reacting phenol and tert-butyl chloride with excess alkali in alcohol. However, commercially the most popular process is the reaction of phenol with isobutylene in presence of 98% H_2SO_4.

Solubility: Practically insoluble in cold water, soluble in alcohol, ether and some other popular organic solvents.

Para-tert-butyl phenol is a classical example of hindered phenols, a major group of antioxidants. They are presently used as radical trapping agents in plastics, elastomers, synthetic fibers, fuels, lubricants and in foods and animal feeds. Is also used as a soap antioxidant, intermediate in de-emulsifiers for oil field and motor oil additives.

3.3 6-tert-butyl-meta cresol

Other synonyms: 6-tert-butyl-3 methyl phenol
2-tert-butyl-6 methyl phenol
Mono-butyl-meta cresol (MBMC)

CAS No. [88-60-8]	
mol. wt.	164.25
d_4^{20}	0.922
mol. formula	$(CH_3)_3CC_6H_3(OH)\ CH_3$ or $C_{11}H_{16}O$
Chemical structure	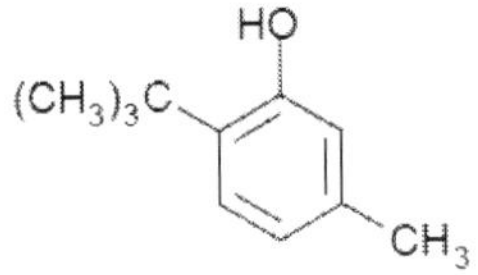
	Clear liquid Solidifies slightly below room temp.

freezing point	23.1° C
bp	244° C

6-tert-butyl-3 methyl phenol also known as monobutyl-meta cresol is obtained as a co-product during Alkylation / Butylation of meta-paracresol for production of Butylated hydroxy toluene (BHT) which is discussed subsequently. On stringent alkylation of meta-para cresol with isobutylene, where the main product is BHT, both mono-butyl-meta-cresol and di-isobutyl- meta-cresols are obtained as coproducts, which can be recovered from the mother liquor after crystallization and centrifuging for isolation of BHT. mono- butyl-meta cresol is also a building block for a few down stream derivatives. Both mono-butyl-meta cresol and Di-tert-butyl-meta cresols are used as antioxidants. Commercially these are de-alkylated to produce pure (98%+) meta cresol. Similarly pure meta cresol can be alkylated with iso butylene to produce mono-butyl meta cresol. Catalyst for alkylation is 98% H_2SO_4.

MBMC is soluble in organic solvents and aqueous potassium hydroxide. It is irritant to skin and has moderate fire risk.

Broadly uses of MBMC are

- Either as an antioxidant or for synthesis of downstream antioxidants
- As a germicide and disinfectant.
- As an additive to lubricating oils.
- As a building block for other chemicals, predominantly for fixative perfumes like musk Ambrette, and rubber processing chemicals.

Butylated meta cresols are never as effective antioxidant as Butylated para cresols (Butylated hydroxy toluene or BHT) and uses are limited. However, cost-wise they are cheaper than BHT. As a low-cost antioxidant for some applications, MBMC finds itself attractive.

3.4 Butylated Hydroxy Toluene (BHT)

Also known as di-iso-butyl-para-cresol (DBPC) 2,6-di-tert-butyl-4-methyl phenol.

CAS No. [128-37-0]	
mol. wt	220.34
mp	69° C
mol. formula	$C_{15}H_{24}O$
bp	220° C
Chemical structure of BHT	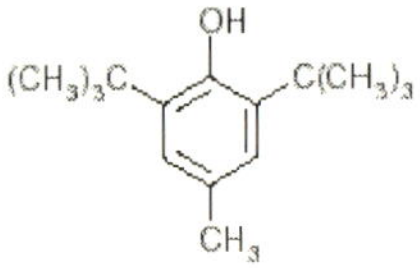
d_4^{20}	1.048
η_D^{75}	1.4859
	White crystals

Solubility: Freely soluble in toluene, soluble in methanol, ethanol, isopropanol, methyl ethyl ketone, acetone, petroleum ether, benzene and in most other hydrocarbon solvents.

A number of antioxidants have been produced and even commercially exploited based on individual cresols and even xylenols isomers. However, para cresol occupies a pride of place for manufacture of antioxidants. Butylated hydroxy toluene (BHT) is the most widely used antioxidant derived from any cresol, phenol or xylenols. [9]

Some of the salient features of BHT are as follows:

- BHT is the single largest volume antioxidant used in rubber, plastics, polyolefins such as, PE, PP, polystyrene and also in gasoline, lubricating oil, polyesters, PVC etc.
- The volatility of BHT is relatively high and hence it is predominantly used in products with lower processing temperatures.
- BHT being a low cost product for many end uses it is the preferred antioxidant.

Production of BHT has started declining. Out of three US producers, PMC, Uniroyal and Merisol, PMC has stopped cresols production and subsequently closed down BHT facilities. It is reliably learnt that smaller manufacturers in Europe and Japan have also closed down BHT operations. On the other hand China has increased production of BHT. Most of the Indian manufacturers in the small scale have also closed down. India now imports good quantity of BHT from Japan. [9]

Worldwide BHT production facilities are shown in Table 3.1.

Table 3.1 World wide BHT Facilities

Company	Raw material	Capacity (tpa)
Merisol, USA	Captive meta-para cresol / para cresol	5500
Uniroyal, USA	Purchased p-cresol	2500
Laporte, UK	Captive p-cresol	5000
Bayer AG, Germany	Captive m-p, cresols	7500
Sumitomo Chemicals, Japan	Captive m-p, cresols	10 to 12000
Toshitomi, Japan	Purchased p-cresol	3500
Mitsui, Japan	Captive m-p-cresol	7500
Yen Shen, China	Captive m-p, cresol	5000 – 7500
Russia	Phenol	8 – 10000
Mexico, Brazil etc.	p-cresol, m-p-cresol	7 – 10000

Considering that USA and some European countries have either banned or restricted uses of BHT for food products, there is hardly and growth in demand in these countries. However, since China, India and some countries in Far East and Africa have expanding petroleum processing and petrochemicals manufacturing facilities it is not unlikely that one or two new BHT plants will be operational. Global demand of BHT has been projected to the tune of some 80,000 tpa.

It is indeed a paradox that on one hand old BHT plants are closing down and new plants are being established in other parts of the world. Even new technologies are getting commercialized. For instance Russia has been operating a phenol based BHT plant for quite some time. Most of the new plants prefer to use meta-para cresol and not pure para cresol for economic reasons. Based on meta-para cresol butylation, plants produce both pure meta-cresol and BHT – both are of commercial importance.

Fig.3.2 gives brief details of production of p-cresol, BHT and pure m-cresol. Fig.3.3 illustrates wide applications of BHT.

By a process of alkylation with isobutylene in presence of an acid catalyst (usually H_2SO_4) gives BHT, DBMC and MBMC. After separation of BHT, the mixture of DBMC and MBMC are dealkylated (using H_2SO_4 at a high temperature) thereby producing pure m-cresol and isobutylene which is recycled. Wide applications of BHT are shown vide Fig.3.3

Most of the BHT manufacturers now prefer to use m-p-cresols firstly, the process is more attractive as along with BHT, pure meta cresol and/or mbmc (mono butyl meta cresol) are produced as co-products having commercial uses. Secondly, pure p-cresol is a much costlier raw material compared to m-p-cresol. p-cresol + m-cresol + isobutylene $\xrightarrow{H_2SO_4 \text{ catalyst}}$ or, p-TSA catalysts

BHT + DBMC and MBMC = $\xrightarrow[\text{at high temp.}]{H_2SO_4 \text{ catalyst}}$ →meta

(Dibul meta cresol) (monobutyl meta cresol) cresol + iso-butylene ↑ recycled.

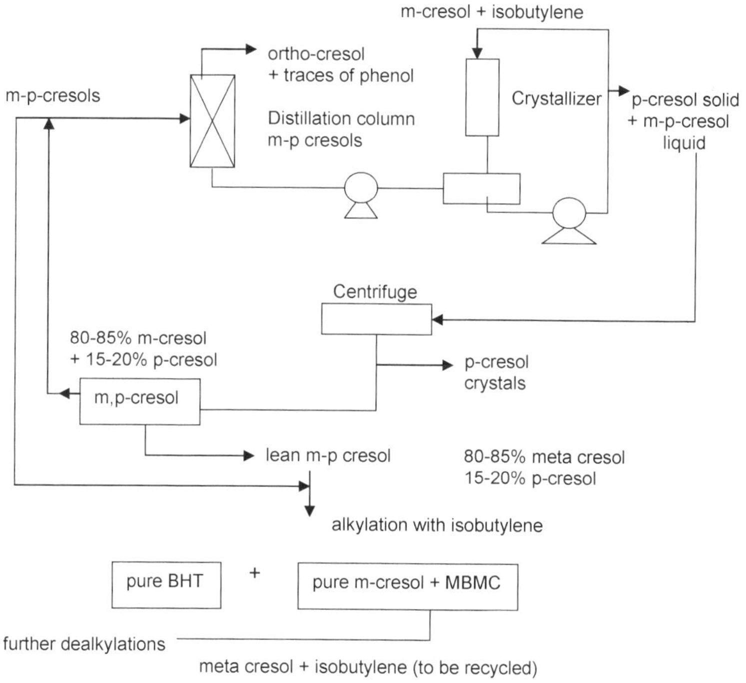

Fig. 3.2: Production of BHT and pure meta-cresol

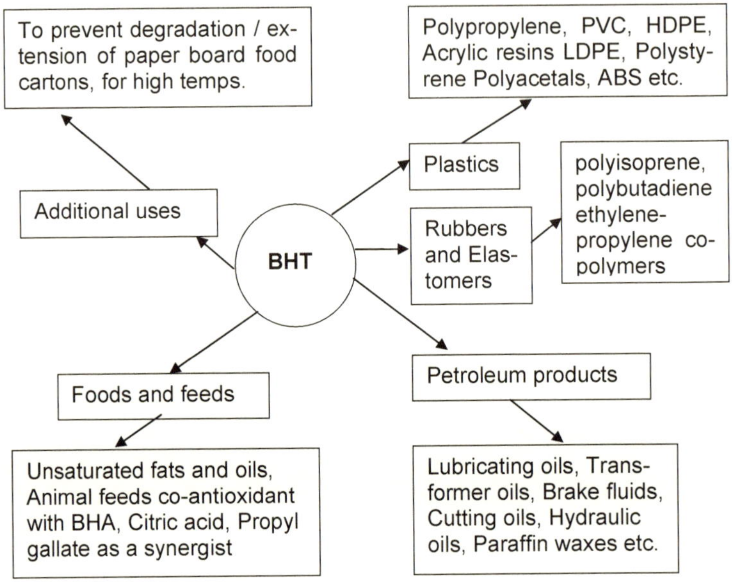

Fig. 3.3 Wide Applications of BHT

3.5 Butylated Hydroxy Anisole (BHA)

Also known as 2- and 3-tert-butyl-4 methoxyphenol, BHA CAS No. [25013-16-5].

BHA CAS No. [25013-16-5]

Chemical structure

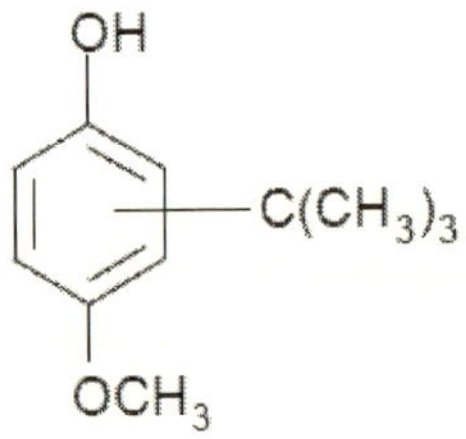

mol. formula	$(CH_3)_3CC_6H_3OH(OCH_3)$ or $C_{11}H_{16}O_2$
mol. wt	180
	White or slightly yellow waxy solid
mp	48 – 55° C
bp_{733}	264 - 270° C

- faint characteristics odor
- Insoluble in water, soluble in petroleum ether, in 50% alcohol or (higher), in propylene glycol, fats and oils
- Exhibits antioxidant properties and synergism with acids, BHT, propyl gallate and hydroquinone etc.

Uses: As an antioxidant and preservative in foods, cosmetics, pharmaceuticals and also rubber and petroleum products. Earlier considered as a replacement of BHT in food products, however, now it has been anticipated as a human carcinogen. It is toxic by ingestion and has restricted uses in food. [1, 4]

BHA in reality is a mixture of two isomeric organic compounds, 2-tert-butyl-4 hydroxy anisole and 3-tert-butyl-4-hydroxy anisole. It is prepared from 4-methoxy phenol and isobutylene. The conjugated aromatic ring of BHA is able to stabilize free radicals sequestering them. By acting as free radical scavengers, further free radical reactions are prevented.

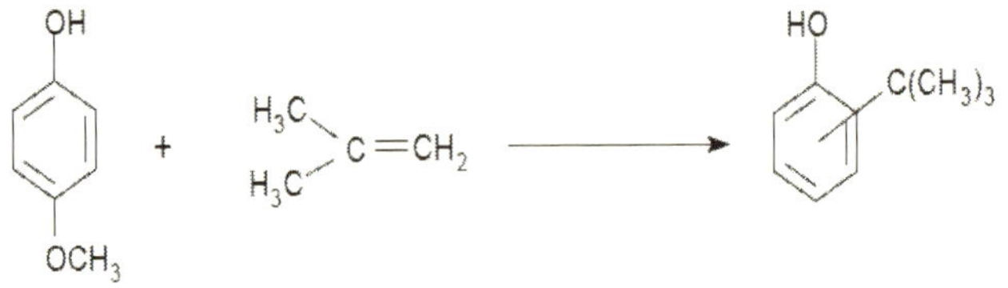

4-methoxy phenol Isobutylene mixture of 2BHA and 3BHA

Commercial food-grade BHA is generally a mixture with >85% 3-tert-butyl-4 hydroxyanisole (3-BHA) and ∠15% 2-tert-butyl-4 hydroxy anisole (2-BHA). Pure food grade BHA is >98% pure. It degrades after prolonged exposure to sunlight. When heated to decomposition BHA emits acrid and irritating fumes.

Cosmetic grade BHA reportedly contains a minimum of 90% 3-BHA and approximately 8% 2 BHA. One product, a lipstick, was reported to contain >10% BHA. Largest users are lipstick manufacturers.

It has been propagated that BHT facilities with some plant modifications might produce BHA. In any case this should be kept in mind that BHT and BHA are two district antioxidants with individual characteristics.

BHA is a synthetic analogue of Vitamin E and operates by reducing oxygen radicals and interrupting the propagation of oxidation processes. BHA can be used in baked products, as it is stable at high temperatures. Primarily used to prevent rancidity in fats and oils.

Typical uses include biscuits, cakes, fats and oils, cereals, pastry and pastry products, sweets etc. in the food products. BHA has been extensively used in petroleum products and petrochemical products.

3.6 TBHQ

Table 3.2 Properties of Hydroquinone and TBHQ

Hydroquinone	**tert-butyl-hydroquinone (TBHQ)**
CAS No. [123-31-9]	CAS No. [1948-33-0]
$C_6H_6O_2$	$C_6H_3(OH)_2C(CH_3)_2$
For. wt. 110.22	For wt. 166.22
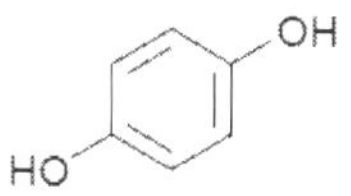 Crystals, mp 170 – 71°C	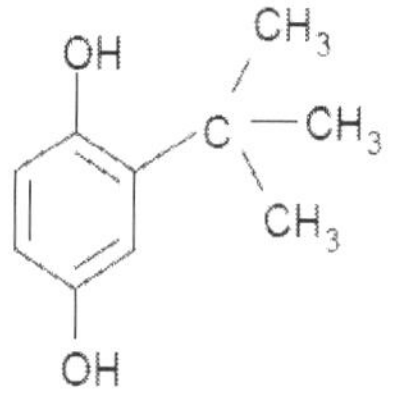 Crystals mp 125°C
Insoluble in water, soluble in alcohol, ether, slightly in benzene	Insoluble in water, soluble in alcohol, acetone and ethyl acetate

Both Hydroquinone (infrequently used) and its butylated derivative tert-butyl-hydroquinone (TBHQ) exhibit antioxidation properties and have been classified as such. TBHQ is a much powerful antioxidant having diversified uses. TBHQ is used as a food antioxidant in USA and some other countries, but not recommended in Europe (EEC) due to lack of adequate toxicological data. Lack of adequate carcinogenicity data on TBHQ is an important gap and it would call for additional research work to generate more authentic data from toxicological and carcinogenic points of view. However, it can be freely used for other products.

TBHQ was accepted as a food antioxidant in a number of countries like USA at a time when there was not much concern about toxicity in general. More recent studies have indicated that

TBHQ may be mutagenic in vivo. However, this will call for detailed investigations and development of additional toxicological data. Unlike BHT/BHA, TBHQ is not supposed to have significant effects on the liver at high doses. However, there are no long-term studies to establish about total risk-free use of TBHQ vis-a-vis other similar antioxidants. Perhaps this is the reason why TBHQ as food antioxidant is still not accepted in Europe and elsewhere. [3]

3.7 Propyl Gallate

Also known as gallic acid propyl ester, 3,4,5-trihydroxy benzoic acid propyl ester or PG.

CAS No. [121-79-9]	
mol. formula	$C_{10}H_{12}O_5$
mol. wt.	212.20

Physically PG is crystalline having mp of 150° C.

Solubility: Soluble in alcohol, petroleum ether

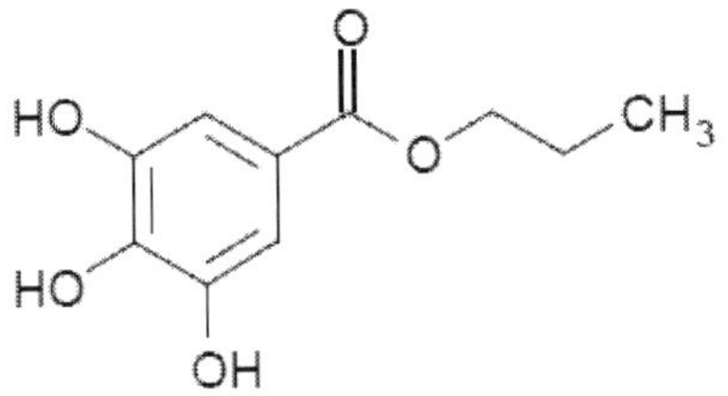

Produced from gallic acid i.e., 3,4,5-trihydroxy benzoic acid

CAS No. [149-91-7]	
mol. formula	$C_7H_6O_5$
mol. wt.	170.12

1)

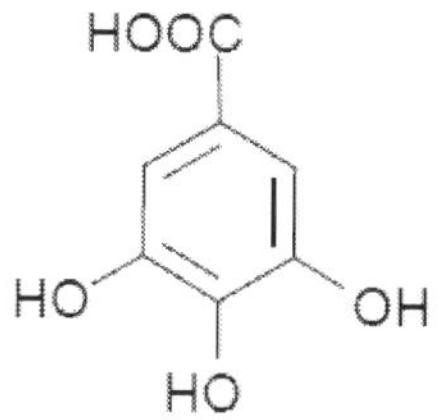

+

2) and propylene $H_2C = CHCH_3$. Propyl Gallate.

Uses: As an antioxidant for cosmetics, foods, fats and oils, ethers, emulsions, waxes, transformer oil. It exhibits synergistic effects with BHT and BHA. Propyl gallate is the most effective antioxidant made from both natural sources as well as it is made synthetically.

Gallic acid is made by hydrolysis of tannins from Tara pods. It exhibits excellent antioxidant activity in food and vegetable oil especially in combination with Ascorbyl palmitate.

Many cosmetic preparations, particularly those containing animals and vegetable oils may deteriorate rapidly due to oxidation. Use of antioxidants delays the process of degradation. Vegetable oils and fats are known to develop rancidity because oxygen combines with them to produce offensive smelling breakdown products. To start with it is slow process but soon it becomes rapid because of a chain reaction. Very small quantities of antioxidants like propyl gallate prevent this rancidity process, 0.1 to 0.5% of the total fat or oil is suggested as the prescribed dose of the antioxidant. PG has greater antioxidant activity than 2-BHA in lard and soybean oil. [20]

Butyl gallate, octyl gallate and Dodecyl gallate along with propyl gallate are very good food antioxidants particularly for protection to lard but they do not 'carry through' to baked goods i.e., they do not give effective protection to fats after it has been incorporated in the food and the product has been cooked.

A second disadvantage of gallates is that in the presence of iron they can form blue-black iron compounds. Other antioxidants like BHT or BHA are also good for protection against oxidative degeneration of products but like gallates does not give good carry through.

Octyl gallate is synthesized by the esterification of gallic acid. It is used as an antioxidant in food, often with BHT and BHA, although it has limited use, as it is unstable at high temperature. Typical products-uses include oils and fats, cereals, snack foods, dairy produce.

Gallic acid, the critical raw material for production of gallates is chiefly produced in China as a natural product from the bark of a particular tree. China exports this product to a number of countries including India. Gallic acid is used for making 3,4,5-trimethoxy-benzaldehyde (TMBA), which is then converted, to Trimethoprim – a popular drug made mostly in India and China. [9]. However, now technology has been developed to make both gallic acid and trimethoxy benzaldehyde form para cresol. [9]

So long Gallic acid or 3,4,5-trihydroxy benzoic acid is easily available, propyl gallate and for that matter, other gallates, will be synthesized and made available to the consumers at a reasonable price. Propanol, the other raw material for making PG is available from petroleum processing units and petrochemical plants (Naphtha or gas crackers). Not only USA, West European countries and

Japan are important producer of propylene and propanol but both India and China produce substantial quantities of propanol / iso-propanol which can be made available after meeting the indigenous demands. It is expected that future researchers will develop more data from toxicity and carcinogenic points of view and establish propyl gallate as a very key antioxidant for different products.

3.8 Vitamin E

This product having highest consumption as an antioxidant has been discussed in some details in the preceding chapter.

Because of increasing demand of Vitamin E, natural sources are not adequate, and in recent years, substantial quantities of Vitamin E or its main constituent α-tocopherol is produced synthetically from meta cresol or 2,6-xylenols. 2,3,6-trimethyl phenol the basic raw material for α-tocopherol is obtained from either m-cresol or 2,6 xylenol. The process was discussed in some details in the author's earlier publication. [9]

Trimethyl phenol is oxidized to trimethyl quinone, which is catalytically reduced to trimethyl hydroquinone, which in turn reacts with phylol to produce α-tocopherol etc.

Phytol, CAS No. [150-86-7], $C_{20}H_{40}O$, an alcohol obtained by the decomposition of chlorophyll, from the plants is an odorless liquid, BP 202 – 204°C/10 mm Hg, has been used in the synthesis of α-tocopherol i.e., Vitamin E. The various possibilities for total chemical synthesis as well as partial synthesis of Vitamin E and particularly α-tocopherol have been extensively described. As told earlier synthesis of α-tocopherol is based on the condensation of 2,3,5-trimethyl hydroquinone with phytol, isophytol has been described earlier. The crude tocopherol is purified by vacuum distillation. Synthetic isopherol used results in a racemic mixture of all eight possible stereoisomers. Following Figure 3.3 briefly gives the chemistry of the process. [3]

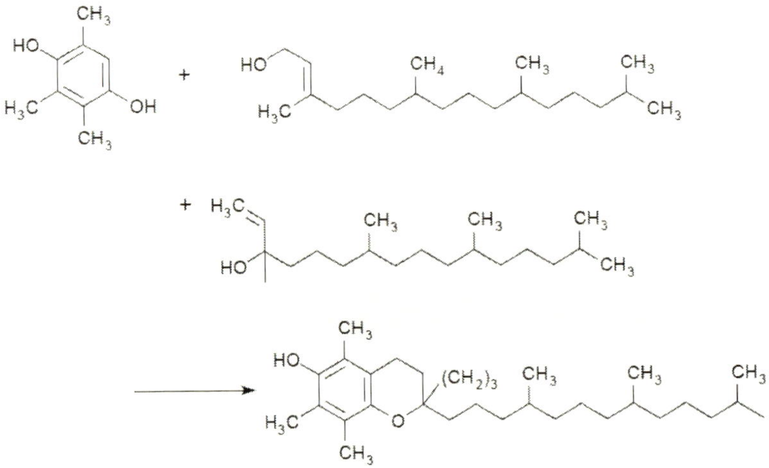

Fig. 3.3 α-tocopherol form 2,3,5-trimethyl-hydroquinone

The fastest growing Vitamin E or α-tocopherol is widely used in diversified fields of food products, fats and oils, pharmaceuticals, cosmetics etc. It is very difficult to predict the demand forecast as it is changing very rapidly. However, a good estimate shows that the total demand for various applications will be to the tune of some 20,000 tpa by the end of 2007.

3.9 Irganox™

A series of hindered phenolic antioxidants were introduced by Ciba-geigy almost 15 years ago. These antioxidants designated as Irganox™ were found to be extremely effective even in small doses and were used as metal deactivators, polyethylene, nylon, polyacetals etc. They have been used also for general polymeric materials, stabilizers for organic and polymeric materials, food packaging, petroleum products and other fields. These were proprietary materials guarded by patent rights and exclusively made and marketed by Ciba-Geigy.

Irganox™ is a series of complex, high molecular wt. stabilizers that inhibit oxidation and thermal degradation of many organic materials. They contain multifunctional chemical groupings, many of them being hindered polyphenols. All are white, crystalline, free flowing powders, non-staining, non-volatile and odorless. Individually each antioxidant in the series was identified by numbers given by Ciba-geigy. Most of them were stated to be soluble in benzene acetone chloroform and some other organic solvents. [21]

Some of the Irganox series of antioxidants are shown vide Table 3.3. [12]

Table 3.3 Some of the Irganox™ series of Antioxidants

Chemical name	CAS registry no.	Trade names
Tetra kis [methylene (3,5-di-tert-butyl-4-hydroxy hydro cinnamate)] methane	[6683-19-8]	Irganox – 1010
--	[41484-35-9]	Irganox – 1035
--	[2082-79-3]	Irganox – 1076
N,N'-1,6-hexamethylene-bis-3- (3,5-di- tert-butyl-4 hydroxy phenyl) propionamide	[23128-74-7]	Irganox – 1098
--	[1709-70-2]	Irganox 1330

Irganox™ 245, 259 and some others have also been introduced by Ciba-geigy. All these antioxidants and stabilizers have been well received in the market.

It is reliably learnt that since most of the patent rights have since expired one or two other chemical companies have introduced or are likely to introduce similar types of products. Sumimoto Chemical Co., Japan, for instance, a leading BHT manufacturer in the field, may be one such company.

3.10 Trihydroxybutyrophenone [a food grade synthetic antioxidant]

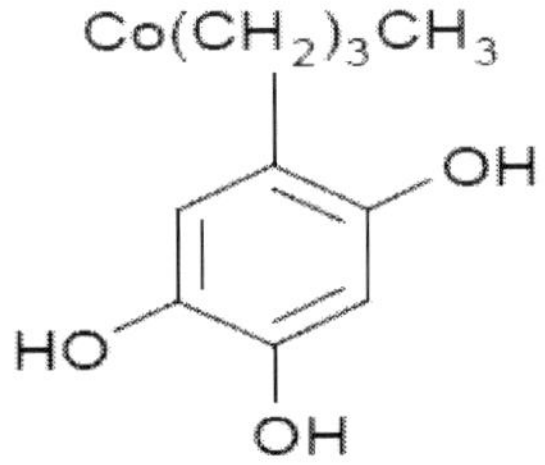

$C_6H_2(OH)_3CO\ C_3H_7.CH_2$

Yellow, tan crystals, mp 149 – 153° C, very slightly soluble in water, soluble in alcohol and propylene glycol. An important antioxidant for polyolefins and paraffin waxes and food additives. Though an antioxidant of recent origin, it is popular in USA. Still not in use in EEC countries [1,3].

3.11 Vitamin C (Ascorbic Acid)

CAS No. [50-81-7]	
mol. formula	$C_6H_8O_6$
mol. wt.	176.12

Vitamin C as a natural antioxidant has already been discussed in some details in the preceding chapter.

However, commercial Vitamin C in substantial quantities is obtained by microbiological and chemical synthesis starting from D-Glucose. The following steps involved in the Synthesis Process is outlined via Fig. 3.4.

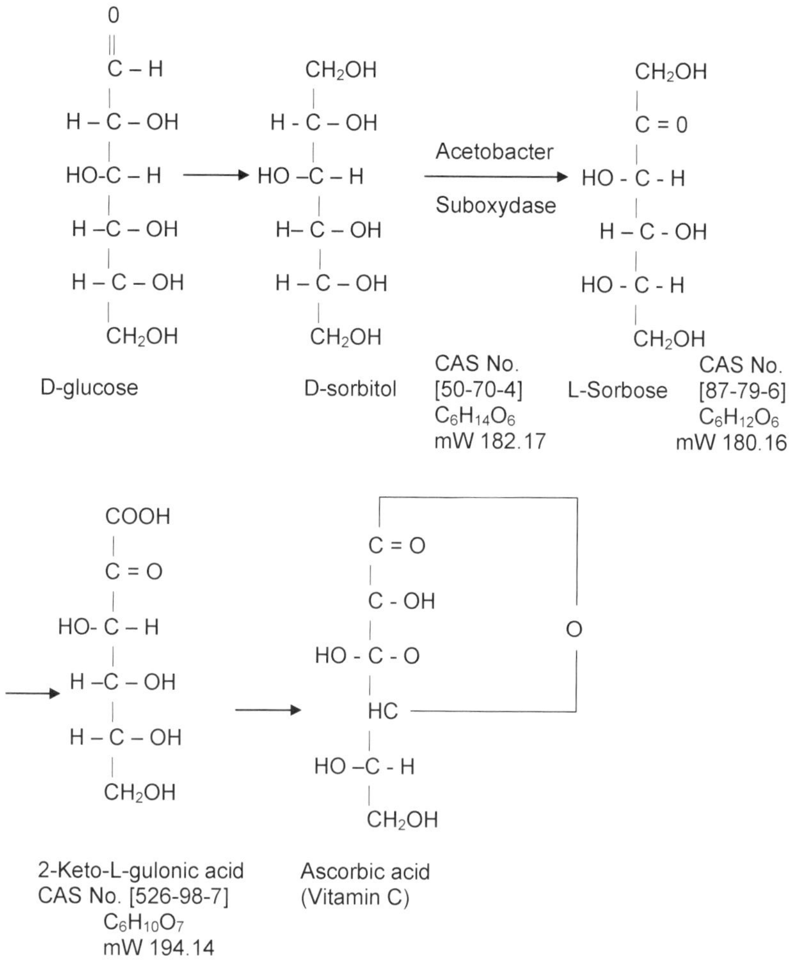

Fig. 3.4 Synthesis of Ascorbic Acid from D-glucose

Sorbitol is made by catalytic hydrogenation of glucose. L-sorbose is produced from sorbitol by the action of several species of bacteria of the genus Acetobacter.

The chemical steps in the conversion of sorbose to ascorbic acid would involve the preparation of the diacetone derivative, which is then oxidized, the acetone groups are removed and the resultant 2-Keto-L-gulonic acid is isomerized to enediol with ring closure.

Attempts have been made to commercialize a single step process using a genetically engineered bacteria to convert glucose to 2-keto-L-gluconic acid directly as Vitamin C precursor and thereafter to Vitamin C.

Vitamin C or ascorbic acid has wide applications in industry as shown vide Figure 3.5.

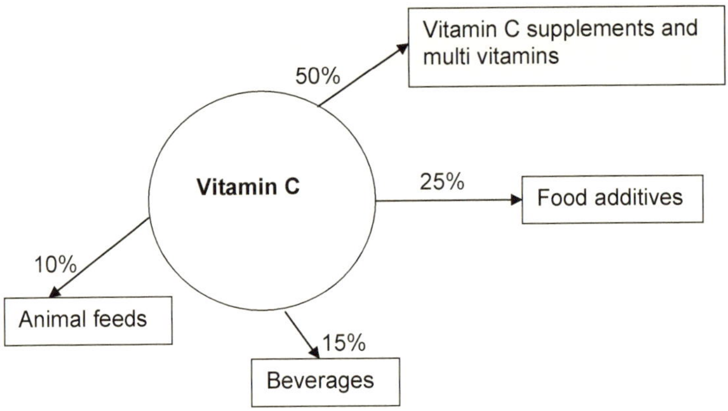

Fig. 3.5 Uses of Vitamin C (ascorbic acid)

The two major USA manufacturers of Vitamin C were Roche with 30 million lbs/yr at its Belvidere, New Jersey facilities and Takeda with 11 million lbs/yr facilities at Wilmington, North Carolina.

The major European manufacturers were Merck and BASF. Merck subsequently closed down its facilities and BASF took over Takeda's bulk vitamins operations in USA in 2000.

China has turned out to be a major Vitamin C (commercial) bulk manufacturer. It is produced in four plants, namely, Northeast Pharmaceutical, North China Pharmaceuticals, Shijiazhong

Pharmaceutical, and Jingshan Pharmaceuticals. Combined production capacity, though not exactly known is stated to be in the vicinity of 70-100,000 tons/yr. [25] [35]

3.12 Ethoxyquin [1, 4, 36]

Also known as 6-ethoxy-1,2-dihydro-2,2,4-trimethyl quinoline.

CAS No. [91-53-2]	
mol. wt	217.31
mol. formula	$C_{14}H_{19}NO$
Chemical structure	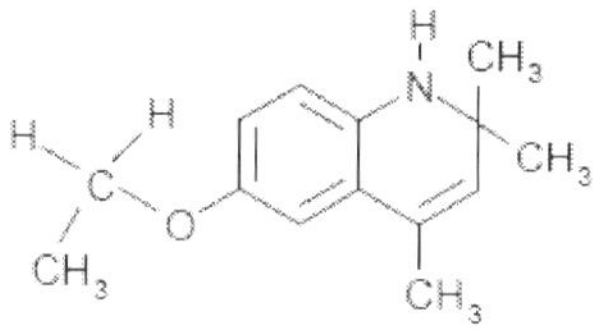
Properties	Yellow liquid
bp	125° C (2 mm Hg)
mp	approximately 0° C
n_D^{25}	1.569 – 1.672
d_{25}^{25}	1.029 – 1.031

Toxic by ingestion. Discolors and stains badly.

Uses:

- Antioxidant in feed and food.
- Anti degradation agent (antioxidant) for rubber.
- Post harvest preservation of apples (scald inhibitor)
- Also used as an insecticide

3.13 Ionox

Trademark (TM) of a series of antioxidants for polymeric systems, thermoplastic rubber and fatty acid distillates, manufactured and marketed by Shell, U.K. [25]

Ionox-100 and Ionox-201 are two classical examples.

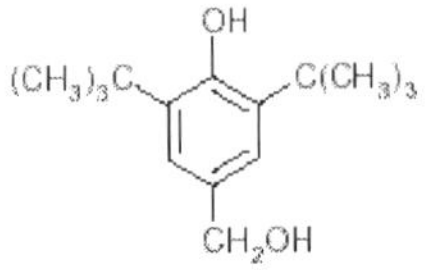

Ionox-100

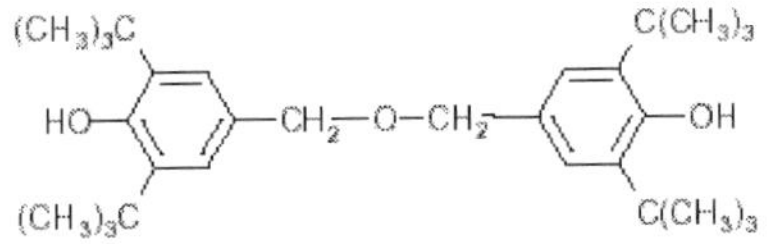

Ionox-201

Synthetic antioxidants were primarily introduced because of expanding horizons of applications of antioxidants. Not all antioxidants, and certainly not all synthetic antioxidants, can be used as inhibitors of oxidative degradation of all products. Many of these synthetic antioxidants were conceived to retard the process of food spoilage. These compounds are believed to act as hydrogen donors or as free radical acceptors that intercept and had quantum of energy that otherwise might induce oxidation. Except trademark antioxidants in the Ionox and Irganox groups most popular food antioxidants are BHT, BHA, TBHQ, PG, THBP etc. 3.6 summarizes these food antioxidants:

Even at the cost of repetition these antioxidants structurally are as shown in Fig.3.7.

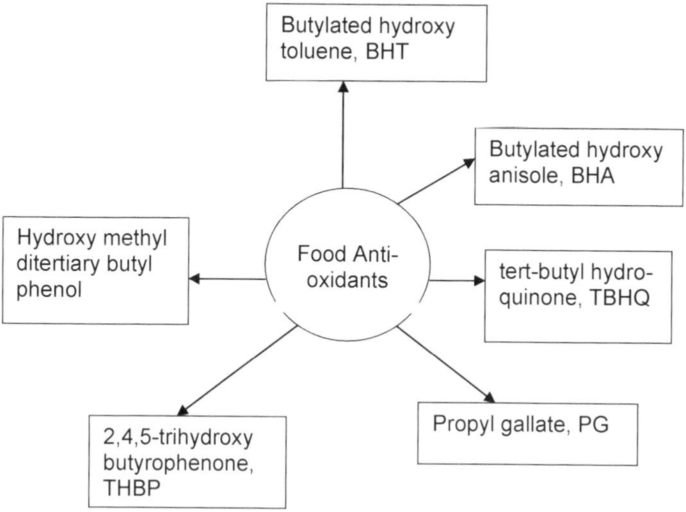

Fig. 3.6 Synthetic Food Antioxidants

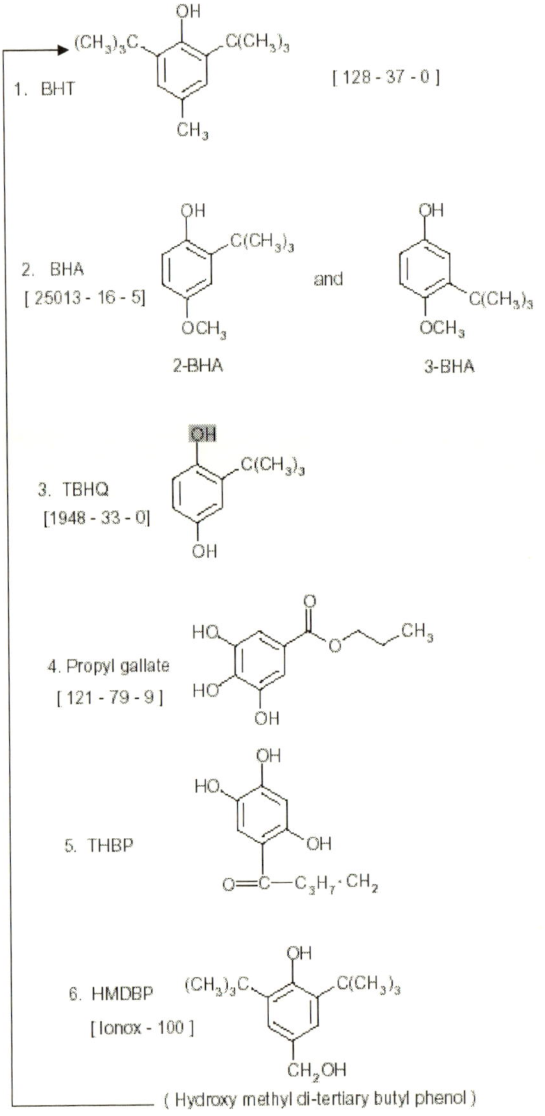

Fig. 3.7 Chemical Structures of Food Antioxidants

Not all these synthetic antioxidants have been approved by the regulatory food and drugs boards of all the countries. However, by and large these antioxidants have been approved by WHO/FAO.

Generally they are allowed in food products at a 0.01% of total fat weight each of 0.02% in combination.

Synergists used in combination with other antioxidants are citric acid, isopropyl acid, phosphoric acid, ascorbic acid, ascorbyl palmitate, erythorbic acid (iso-ascorbic acid), tartaric acid and lecithin, most of which act as chealors of pro-oxidative metals.

Oil of rosemary (containing rosmaridiphenol), extracted from the flowers of rosemars, cultivated in the mediterranean basins with about 1% volatile oils it is a natural food antioxidant.

SELENIUM AND SOME OTHER INORGANIC MICRONUTRIENTS/SEMI-ANTIOXIDANTS

4.1 MICRONUTRIENTS/INORGANIC CHEMICALS

Selenium and some other inorganic elements like zinc; magnesium and sulfur are essential micronutrients playing important roles in human metabolism process. Their absence or presence in less than the RDA would weaken the body and cause various diseases. There are specific fruits and vegetables that contain these elements and it goes without saying that human body or even animals must have a logical ingestion of these inorganic elements. Briefly these are discussed here.

4.2 Selenium

CAS No. [7782-49-2]

A metallic element

Atomic No	34
at wt.	78.96

Amorphous, red powder becoming black on standing and crystalline on heating.

d_4^{16}	4.5
mp	217° C
bp	685° C

There is a characteristic unique biochemical relationship in animals including humans between the element selenium and the fat soluble Vitamin E (α-tocopherol), and the function of interrelationship is highly complex. Selenium confers protection in part by inducing or activating cellular free radical scavenging systems while Vitamin E acts by peroxide breakdown. There is definite synergism between Selenium and Vitamin E. It has been somewhat concluded that all selenium deficiency diseases in humans that can be at least partially prevented by Vitamin E, although whether Vitamin E in all cases can be quantitatively replaced by selenium is not established.

It must be emphasized that selenium is not an antioxidant. In many species selenium requirements can be met by increases in dietary methionine or cysteine. Then the question arises as to whether selenium is a micronutrient or is it something like a semi-antioxidant.

Selenium requirements are not very significant and considering the toxicity of selenium salts, supplementation in affected areas must be cautiously carried out. Oral selenium is less effective, but heavy selenium pellets in the system have been used successfully.

Selenium is a very rare element. The metal does not occur in nature in free elemental form. Its abundance in the earth's crust is about 0.05 mg/kg. Although a toxic metal, in trace amount is a nutritional element as mentioned above. Trace amounts added to cattle feed are effective against muscular dystrophy in sheep and cattle. It has been stated that in spite of being an essential nutrient at trace concentrations, selenium is highly toxic at moderate concentrations. Exposure to selenium metal fumes can cause severe irritation of eyes, nose and throat. [22]

- Selenium exhibits synergistic activity in presence of Vitamin E.

- Selenium protects against cancer particularly lung, prostate, and colon cancer. Inadequate selenium content in some soils may result in stroke and heart disease in humans.
- Selenium is not produced by the body and is procured through food and water. Food sources of selenium include garlic, onions, wheat germs, red grapes, broccoli and egg yolks.
- The RDA for selenium is 75 micrograms for men and 55 micrograms for women. Not consuming sufficient quantity of selenium within the prescribed limits could be dangerous.

Canada, Japan, Mexico and certain soils in erstwhile Yugoslavia are the chief sources of selenium. The selenium deficient states in USA include Connecticut, Illinois, Ohio, Oregon, Massachusetts, Rhode Island, New York, Pennsylvania, Indiana and Delaware as well as the District of Columbia. People living in these states are definitely in need of selenium supplements.

Same is true for most African, Asian and European countries.

4.3 Zinc, Magnesium and Sulfur [22]

4.3.1 Zinc

CAS No. [7440-66-6], at wt. 65.39, is an essential micronutrient for many species and it is certainly required by man. Zinc is believed to play an important role in normal wounds in man and other animals. The main function of zinc in the body is believed to be as a co-factor for nucleic acid protein synthesis, and in carbohydrate metabolism.

The richest dietary sources are wheat germ (50-100 ppm) and certain shellfish. Cereals, vegetables, meat, eggs, fish and legumes all contain considerable amounts of zinc. The daily ingestion of a human body is 5-15 mg/day.

Again it is to be re-emphasized that zinc is a micronutrient and not an antioxidant, though inadequate zinc in the human system may cause severe health problem.

4.3.2 Magnesium

CAS No. [7439-95-4], at wt. 24.30.
Magnesium is an essential element for man and other animals being a component of soft tissues and bones. In man the normal plasma has a magnesium level in the range 1.75 – 2.34 milli-equivalents per liter. The adult human body contains 21 – 28 gm of magnesium.

Dietary sources of magnesium include Flour, Wheat (120 mg/100 gm), Oatmeal (150 mg/100 gm), Soy flour (220 mg/100 gm), Cocoa (400 mg/100 gm), Peanuts (170 mg/100 gm), Beans (180 mg/100 gm), Spinach (60 mg/100 gm) etc.

As in ease of zinc, magnesium is a micronutrient and is considered essential for the human body. Adequate magnesium will be available by intake of the above food-products. It is not an anti-oxidant but is considered a must for a healthy body.

4.3.3 Sulfur

CAS No. [7704-34-9], at wt. 32.066, is one of the components of all living tissues, being essential for the basic building units of structures such as muscle and cartilage and playing a ubiquitous physiological role in cellular metabolism.

Sulfur is present in all vegetable foodstuffs, especially in the structural tissues of plants; sulfur content in common rich vegetable sources include white bread (80 mg/100 gm), Oatmeal (150 mg/100 gm), Beans (100-150 mg/100 gm), Cabbage (80 mg/100 gm), Onions (170 mg/100 gm), Peas (130 mg/100 gm), Peanuts (300 mg/100 gm) etc.

Out of the four micronutrients discussed above, selenium has a pride of place. Selenium is not only a micronutrient but has proven semi-antioxidant properties as selenium has synergistic activities with Vitamin E. It is thus a special micronutrient and a lot of research work is going on in the field of selenium.

Other three elements zinc, magnesium and sulfur, and may be even Mn, as of date have only utilities as micronutrients. For a healthy body all these micronutrients are necessary. In their absence or inadequate intake of such micronutrients from different foodstuffs, various diseases, many of them serious, may occur and medical complications including premature aging may take place. Selenium is, as of date, not classified as an antioxidant though it does catalyse the boosting up of the functional effectiveness of Vitamin E, a tocopherol of great importance, to the system. Other three elements might have also special functions for boosting the immunity of the system. Though this is a new terminology it may be logical to call these groups of elements as semi-antioxidants.

As of to-day antioxidants, natural or synthetic, are all organic chemicals. There is nothing like inorganic antioxidants. Mineral or inorganic based chemicals are micronutrients essential for the animal bodies. Who knows future researchers may come to a conclusion that these inorganic chemicals, which boost up the immunity systems and in some cases even aging and in cxtreme cases premature deaths? So why not classify them as semi-antioxidants?

5

EXTRACTION OF ANTIOXIDANTS

5.1 BACKDROP

Spices, herbs, fruits, vegetable and tree materials (roots, bark, leaves etc.) are used not only to enhance flavor and/or as nutrients and antioxidants containing phenolic and polyphenolic materials, are extracted from the original mother plants and herbs. The extraction processes commercially employed for the effective antioxidant in highly concentrated form include:

i) extraction with both polar and non-polar solvents e.g., ethanol, ethyl acetate, n-hexane, acetone, methyl ethyl ketone, dichloromethane, etc.,
ii) extraction with aqueous NaOH/KOH solutions,
iii) extraction with vegetable oils and/or monoglycerides or diglycerides,
iv) Steam distillation,
v) molecular distillation, and
vi) during last two decades the supercritical CO_2 extraction.

The main drawback of liquid-liquid solvent extraction has limitations, as the solvents used are not sufficiently selective for the active antioxidant ingredients. Molecular distillation suffers from the problem that there might be dilution with the carrier used in the distillation process. Similarly there might be an adverse effect on their solubilization with fats and oils.

Supercritical fluid extraction using CO_2, on the other hand, overcomes most of the above-mentioned drawbacks and prima-facie looks like a very attractive method. Only solvent extraction and supercritical fluid (SCFE) extraction using an inert gas like CO_2 are described here in some details.

5.2 Detection and Extraction of Antioxidants

Natural tocopherols (Vitamin E), ascorbic acid (Vitamin C) and common synthetic antioxidants (BHT, BHA and TBHQ etc.) in oils, fats and fatty foods have been studied for their detection and determination and there after extraction. Numerous methods, more popular ones being spectrophotometry, thin-layer chromatography, gas-liquid chromatography (GLC) and high performance liquid chromatography (HPLC) have been described in literature for the qualitative and quantitative determination of antioxidants in food and other products.

5.2.1 Extraction of Antioxidants

A lot of work has been done and reported in case of extraction of antioxidants using suitable solvents. For instance two common vegetables like Tomato and Carrot are rich source of carotenoids lycopene and β-carotene. Similarly marigold flower is a rich source of lutein, another carotenoid. These are also sources of neutraceuticals, which can be extracted from the mother vegetables both by solvent extraction as well as by supercritical fluid extraction (SCFE) using high pressure CO_2.

Lycopene and β-carotene are isomers and lutein is the oxidized form of lycopene and/or β-carotene though lutein is also a natural antioxidant and cannot be made form lycopene or any of the four carotenes (α, β, γ- and δ). Lycopene, β-carotene and lutein have good solubility in organic solvents such as Dichloro methane, (also known as methylene chloride). Benzene, CCl_4, Petroleum ether, chloroform etc.

Not published data and information are available regarding recovery of these products via solvent extraction. There are a number of European patents WO 03079816, EP 1384472, CN 1198661, CN 1298904, WO 0179355, EP 0818225, WO 03103645, WO 03103645, US Patent 5837311 etc. [24]

Taking an example of lycopene extraction from tomato paste it is usually suggested that neutraceutical concentrate via solvent extraction suffers from a few drawbacks such as i) Poor extraction and selectivity of separation of the active ingredient, ii) Presence of traces of undesirable solvents and consequent difficulties in high temperature processing needed for removal of residual solvent, iii) requirement of multiple solvents and consequently multiple solvent recovery steps.

Extraction of β-carotene from carrots or lutein from marigold flowers using same solvent treatment process is virtually identical with solvent extraction of lycopene from tomato paste. In some cases n-hexane has been used as one of the solvents for solvent extraction process.

A typical carotenoid like β-carotene has been extracted from fresh carrots using water-miscible solvents, usually, acetone, methanol or ethanol. Although cold solvent may be used to minimize loss of the active ingredient like β-carotene or even lutein from marigold flowers. A hot solvent i.e., relatively high temperature facilitates better extraction. Soxhlet apparatus, in a laboratory or bench-scale is ideal for extraction of antioxidants, which are extracted and come out along with some amount of solvent. Thereafter extracted material is recovered using vacuum distillation. After drying and grinding enriched feed (i.e., separated antioxidant) is ideally suited (94-95% pure) for application or use for any food product, which needs an antioxidant for preservation, storage or packaging. Sometimes enriched feed is re-extracted with a solvent like methanol. In an industrial plant same principles are adopted for recovery of the antioxidant in a much larger scale.

Broadly recovery methods using solvent extraction (use of 2 solvents in different steps is common) is shown vide Figure 5.1. Lycopene extraction from fresh tomatoes or tomato paste is a typical example. [14]

Recovery of an antioxidant like Lycopene is a typical case. However, if very pure lycopene is required (> 99% pure). Supercritical fluid extraction (SCFE) of the enriched feed is an ideal solution. An SCFE set-up is shown vide Fig.5.2.

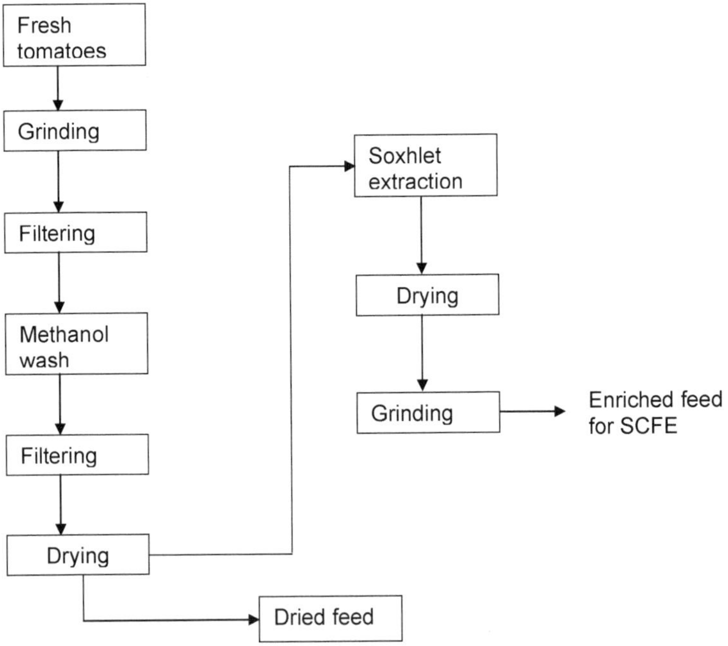

Fig. 5.1 Recovery of Antioxidant (Lycopene)

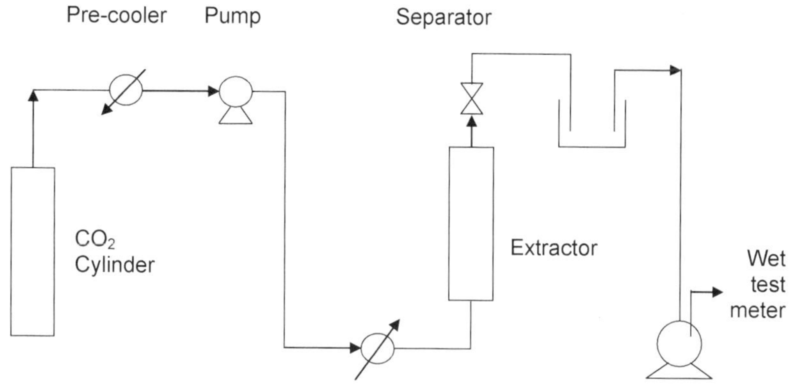

Fig. 5.2 SCFE Set up using High Pressure CO_2

[Pressure 300 – 450 bar
Temp. 50 – 80° C
Time 3 – 5 hrs]

Fig. 5.3 shows the schematic description of feed pre-treatment and SCF fractionation.

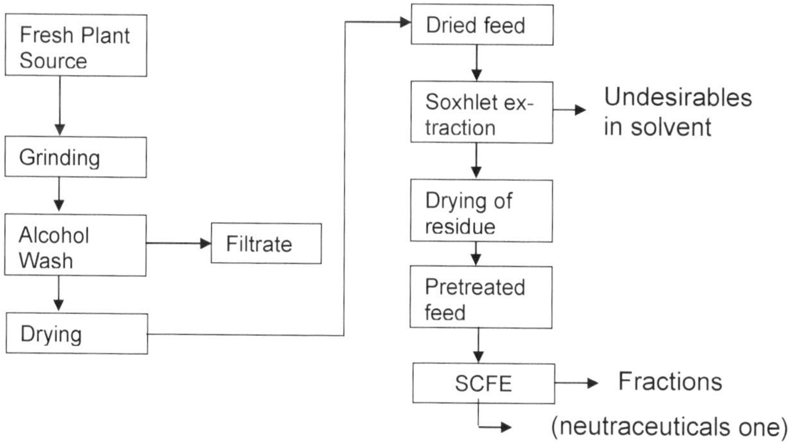

Fig. 5.3 Schematic block diagram for feed pre-treatment and SCF fractionation

Use of SCFE using high pressure CO_2 needs a critical review. After use of two solvents for conventional solvent extraction purity of the recovered product, antioxidant, lycopene from tomatoes, β-carotene from carrots or lutein from marigold flowerer or any other antioxidant from natural sources or if synthetically made, purity of the antioxidant will be in the vicinity of 90-95%. SCFE will be capital intensive being a high-pressure process, and considering additional operating cost, it will strictly call for a cost-benefit analysis before introduction of another extraction step namely SCFE. Purity of the final recovered product i.e., very pure antioxidant (say 99% + pure) needs very serious scrutiny about necessity for use of such high pure material. A commercial grade of purity (90-95%) may be perhaps adequate.

5.3 Antioxidant Activity of Plant Extracts Containing Phenolics

Phenolic compounds are commonly found in both edible and non-edible plants and they have very significant antioxidant activity. Crude extracts of fruits, herbs, vegetables and other plant materials rich in phenolics are assuming great interest in the food industry because they inhibit the process of oxidative degeneration of lipids. Consequently the quality and nutritional value of food is improved. In an article published on the subject, Professor Marja P. Kahkonen and his colleagues and coworkers of the Department of Applied Chemistry and Microbiology, Division of Food Chemistry, University of Helsinki and Department of Pharmacy, Division of Pharmacognosy, University of Helsinki and Department of Chemistry, University of Turku, all of Finland, have elaborated their studies and research work in the Journal of Agricultural Food Chemistry, in 1999. [23]

Different Berries and fruits, vegetables, Herbs and medicinal plants, cereals and tree materials were examined by the authors. The total phenolic contents were expressed gallic acid equivalents (GAE) in mg/gm of dried material.

Antioxidant testing was carried out by oxidizing Methyl linoleate (MeLo) purchased from a standard source, in the presence of antioxidants. Results of these experiments and investigations are shown vide Table 5.1. Only some selected items encompassing Berries and fruits, vegetables and tree materials have been shown. In the original paper 92 different plant materials including herbs and medicinal plant and cereals were listed.

Table 5.1 Total Phenolics in Some Selected Plant Extracts

Raw materials	drying method a	extraction method b	dry wt. of extract (mg)	total phenolics @ (mg of GAE / gm)
Berries and Fruits Crowberry Empetrum nigram	1	A	71	50.8±1.0 (98%)
Strawberry Fragaria ananassa	1	A	42	14.8±0.3 (57%)
Goosberry Ribes grossularia	1	A	50	12.4±0.6 (93%)
Black current Ribes nigram	1	A	67	20.3±0.6 (93%)
Apple Malus pumila	1	A	26	12.1±0.3(98%)
Rose fruit, Rose sp	1	B	37	12.5±0.1 (12%)
Vegetables Onion Allium cepa	2	B	36	3.0±0.1 (12%)
Red onion Allium cepa	2	B	36	3.0±0.1 (12%)
Beet root, peel, Beta vulgaris esculation	1	C	236	4.3±0.2 (98%)
Carrot, flesh, Daucus crota	1	C	281	0.6 ± 0.0 (0%)
Tomato, hycopersicum esculentum	1	A	21	2.0±0.1 (52%)
Pea legume, pisum savitum	1	C	177	1.6±0.1 (24%)
Potato, peel, solanum tuberosum	1	C	67	4.3±0.2 (66%)
Tree materials Maple leaf, Aces platanoids	1	D	151	31.7±0.2 (65%)
Silver birch, leaf, Betula pendula	1	D	219	38.4±1.0 (63%)
Scotch pine, bark pinus sylvestris	1	C	132	76.0±2.9 (90%)

[[a] 1 = air drying: 2 = lycophilization b extraction procedures A = 70% aqueous acetone B = 80% aqueous methanol, sonication C = 80% aqueous methanol D = 70% aqueous acetone, magnetic mixer C mean of duplicate assays d_w = dry wt. of the original samples.]

Obviously total phenolic content does not give a full picture of the quantity and quality of the phenolic constituents in the extracts. Some interference from other chemical components was also there: like sugar and ascorbic acid.

A major portion of the antioxidant activity of oil-seeds and oil-seed flours and concentrates is attributable to flavonids and hydroxylated cinnamic acids. The combined influence of these isolated phenolic compounds accounts for nearly the antioxidant activity of Soy, cottonseeds etc. flours and concentrates. Similar is the case with Peanuts and peanut flour.

Gallic acid and its esters are of course well known as potent antioxidants. This also shows that cinnamic acid skeleton, as distinct from that of benzoic acid, is not essential for activity. [3]

6

Oxidative Degeneration of Phenolic Antioxidants

6.1 DEGRADATION OF ANTIOXIDANTS

It is indeed a paradox that antioxidants, which were primarily meant for inhibition or retardation of food and other products including fats and oils, are themselves oxidized after a reasonable period of storage and packaging. Since most of the antioxidants – both natural and synthetic are primarily hindered phenols, Diphenols and other polyphenols it will be interesting and important to study the implications and chemistry of degradation of phenolic antioxidants.

Fats and oils are good examples. Even after addition of a suitable dose of antioxidants they become rancid i.e., degraded unless they are kept in refrigerators. Auto-oxidation of fats and oils progresses through a chain reaction of free radicals. Many studies on the mechanism of the action of the antioxidants have been made. It was shown that the antioxidants scavenge free radicals produced during the propagation steps of auto-oxidation of fats and oils by donating hydrogen radicals. Against this background it becomes all the more necessary to understand and study the chemistry of the degradation of antioxidant molecules during the course of auto-oxidation of fats and oils for not only establishing the action of the antioxidants and of synergistic effects of compounds that potentiate their action.

The capability of fat, oil and fatty food to retain a fresh taste and odor during storage and use is usually referred to its stability. It is related to the ester linkages of triglycerides, which are earlier subjected to hydrolysis liberating free fatty acids (lipolysis) or may undergo oxidation forming peroxides and aldehydes (oxidation). Lipolysis can be a result of chemical, /enzymatic reaction or by a thermal stress. Lipolytic spoilage is known as lipolytic rancidity or hydrolytic rancidity. Fats and oils also can become rancid as a result of oxidation and this causes oxidative rancidity.

Lipolytic rancidity usually poses less of a flavor problem than oxidative rancidity because the former develops off-flavor on those fats containing less than C_{12} fatty acids. Oxidative rancidity of because the former develops off-flavor on those fats containing less than C_{12} fatty acids. Oxidative rancidity of fats and oils is thought to be the major cause of deterioration because of ester linkage of polyunsaturated oils. [25]

6.2 Major Food Grade Synthetic Antioxidants

6.2.1 BHT, BHA, PG, TBHQ and THBP: Degeneration

The concentration of an antioxidant in food fats is important for reasons of cost, safety, sensory properties and functionality as they are allowed in food products at 0.01 percent level. Citric acid, isopropyl acid, phosphoric acid, ascorbic acid and tartaric acid are sometimes added as synergists to enhance the effectiveness of antioxidants. [25]

BHT has been used as a very common antioxidant in the group of phenolic antioxidants. Degradation of phenolic antioxidants like BHT was studied in details by Japanese researchers in the Tokyo College of Pharmacy, Tokyo, Japan. Their observations and results were published in the journal of Hygienic Chemistry,

Japan in 1967. Results were reported by Harano, Y., Hoshino, O., and Ukita. [26] A detailed review was published in the book Food Antioxidants edited by Mr. B.J.F. Hudson [3]. They first isolated and identified the products of degradation of BHT in auto-oxidized soybean oil. Soybean oil containing BHT was irradiated by UV light, and decomposition products were detected by thin layer and gas chromatography. Decomposition products are shown via Fig.6.1.

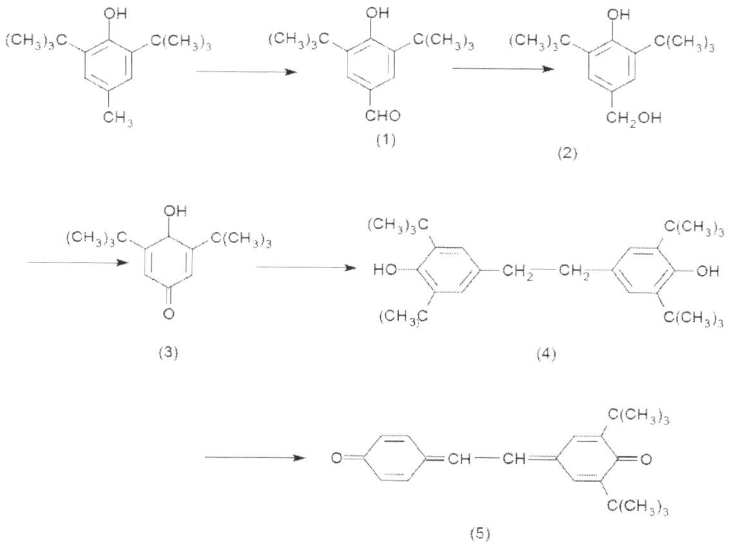

Fig. 6.1 Degradation products of BHT

It was observed that degradation products (1) – (5) were formed as a result of loss of BHT. BHT was completely lost after the induction period of the oil came to an end. The sequences of products are shown above and follow. The following order: Butylated hydroxy toluene → butylated hydroxy aldehyde → Butylated hydroxy butyl alcohol › Butylated hydroxy benzoquinone →

3,5,3',5'-tetra-tert-butyl-4,4'-dihydroxy – 1,2-diphenyl-ethylene → 3,5,3',5'-tetra-tert-butyl still benoquinone.

Butylated hydroxy anisole (BHA) used, as a food antioxidant is a 9:1 mixture of 2-BHA and 3-BHA. Degradation of BHA by light or heat shows that UV irradiation of 2-BHA in benzene gives two oxidation products is shown by Fig. 6.2:

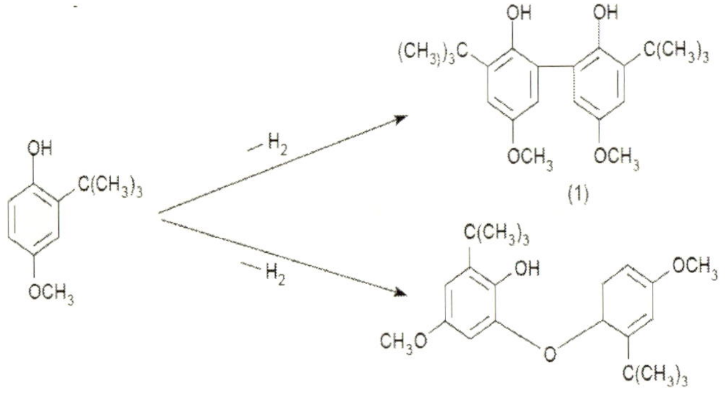

(1)

(2)

Fig. 6.2 Oxidation of Food Antioxidant BHA

Products of oxidative degradation of 2-BHA give two oxidation products, which were detected 2-BHA originally used as the antioxidants. Both compounds were reported to have been formed and 2-BHA was completely lost, (1) and (2) new products formed are:

(1) A biphenyl type dimer, 2,2'-dihydroxy-5,5'- dimethoxy-3,3'-di-tert-butyl biphenol.

(2) A biphenyl ether type dimer 2',-3-di-tert-butyl-2-hydroxy-4,5-dimethoxy biphenyl ether.

Irradiation of a mixture of BHT and BHA in benzene by ultraviolet light affords a new product given below:

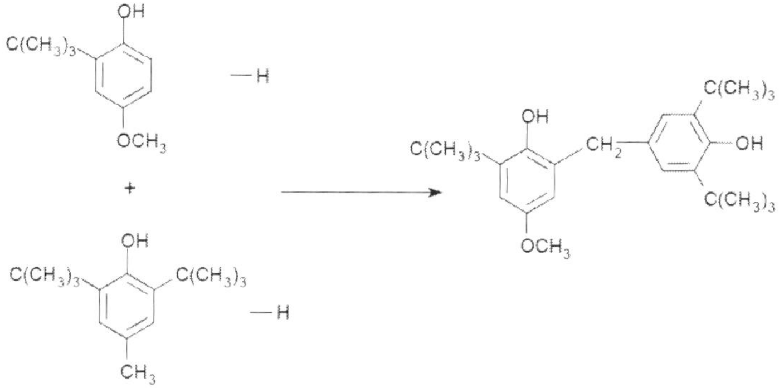

3,3',5'-tri-tert-butyl-5-methoxy-2,4'-dihydroxy diphenyl methane, a bi-pheyl ether type dimer, which is presumably derived from radicals of BHT and 2-BHA.

Similar degradation effects were reported in the synergistic reactions of BHA and PG. A variety of chemical oxidation products of tocopherols have been also reported. [27, 28]

TBHQ another antioxidant used primarily for food products has also been studied for degradation effects in benzene and UV light. For details the reader may refer to studies by Kurechi et al., [29] [30].

Oxidative degradation of THBP, which has recently been introduced as a food antioxidant, primarily in USA has not been studied in details as yet for its oxidative degradation. It is still under investigations.

The antioxidant activities of oxidation products of TBHQ are interesting since the oxidation products still retain antioxidant properties. However, the activity levels are very much dependent on the substrate oils.

Antioxidant properties of oxidation products of TBHQ in methyl oleate, soybean oil and lard oxidized by active oxygen method were studied and reported by Kurechi, T. et al., in J. American Chemical Society. [30]

In sum, detailed investigative studies by some Japanese Scientists and others would confirm the popular belief that after some time at least food antioxidants, so far studied, BHT, BHA, PG, TBHQ, would show degeneration effects, and antioxidant properties after a reasonable period would be lost, and the food products would be rancid, toxic and even develop carcinogenic characteristics. Increasing the antioxidant dose will not be a solution because to start with, the said food products (fats, oils or other food products) would be toxic and reactions cannot be reversed.

Effects of antioxidants in other products such as petroleum products like gasoline, aviation turbine fuel, lubricating oils, almost all plastic products, polymers and elastomers would be similarly affected by the degraded antioxidants if they are not used up within a reasonable period. Effects of light and heat on petroleum products including petrochemicals, mostly olefins and polyolefins, is well known and antioxidants can only protect these products only for sometime, ultimately oxidative degeneration takes place and the antioxidant itself is converted to degeneration products and becomes ineffective.

Only difference is that in case of food products after a reasonable period say 3-6 months, unless kept in refrigerators, the product being spoiled and rancid has to be just thrown out, and is definitely not fit for consumption, otherwise because of toxicity and carcinogenic effects there will be serious diseases and even deaths in some cases.

In other products, predominantly petroleum and petrochemicals quality is seriously affected if the antioxidant used is chemically degraded and becomes ineffective.

Almost all organic compounds are prone to oxidative degeneration (auto-oxidation or otherwise) and this is primarily the reason why a suitable antioxidant, within the prescribed safety limit, is added to guard against oxidation. But studies have conclusively established that almost all antioxidants have a limited shelf life and they themselves become oxidized and render the product unacceptable after some time.

7

Commercial Implications of Antioxidants

7.1 COMMERCIALIZATION OF ANTIOXIDANTS

More than one hundred antioxidants have been detected, determined and isolated in almost pure forms from natural sources for commercial uses. Many of the extensively used antioxidants have been also synthesized to augment, supplement or replace natural ones. Some of the antioxidants have been exploited commercially. Again some known antioxidants have not been commercially exploited as yet. Most critical applications of antioxidants are in the processing, storage and packages of food products. I

Important reason for being selective to choose a particular antioxidant for food uses is primarily effectiveness of the antioxidant for food uses, on consideration of safety, toxicity and carcinogenicity. Also shelf life of a particular food item after addition of an antioxidant is very important. Processed food and fats and oils have a tendency for spoilage since they develop rancidity within a short span. In other words even an antioxidant added food product develops rancidity after some time. The antioxidant itself becomes inactive and degrades, and products of oxidation may not be of any use and foodstuff becomes spoiled.

7.2 Food Antioxidants

Most commonly used in foods, and relatively more widely used, and more popularly accepted antioxidants are shown vide Tablc 7.1. [31] [3]

Table 7.1 Antioxidants most commonly used in foods

Antioxidant (common abbreviation)	'E' number[a]	Typical uses
L-Ascorbic acid (Vitamin C)	E 300	Fruit juices, drinks mayonnaise, cured meat, fish products, butter etc.
Synthetic α-tocopherol	E 307	Infant foods, milk, fat, mayonnaise
Propyl gallate (PG)	E 310	Chewing gum, vegetable oils
Butylated hydroxy anisole (BHA)	E 320	Animal fat, cheese spread, biscuits, potato flakes, beef stock cubes
Butylated hydroxy toluene (BHT)	E 321	Walnuts, chewing gum
Citric Acid	E 330	Vegetable oils, mayonnaise
Tertiary butyl hydroquinone (TBHQ)[b]		Palm oil, frying oils
2,6-di-tert-butyl-4-hydroxy methyl phenol (Ionox-100), and 2,4,5 Trihydroxy butyrophenone (THBP) are rarely used in EEC.		

'a' E numbers are the European Economic Community (EEC) codes

'b' Not permitted in the EEC countries as of now.

The EEC approach on safety and necessity of food antioxidants has been elaborated by Haigh. [31] Recently WHO/FAO Codex Committee on Fats and Oils recommended the revised maximum levels of food antioxidants as 100, 75, 175 and 120 mg/kg for propyl gallate, BHT, BHA and TBHQ respectively. They further recommended that any combination of these additives should not exceed 200 mg/kg. It was also recommended that for tocopherols the value should be 500 mg/kg maximum, for ascorbyl palmitate and ascorbyl stearate maximum value should be 500 mg/kg. [32]

The list of antioxidants used in industry is impressive. More than 150 antioxidants have been mentioned having uses in diversified fields. The list of antioxidants permitted for food use is much shorter (Table 7.2).

Table 7.2 Usual Food Antioxidants Commercially Exploited

Primary antioxidant compound	Derivatives
Ascorbic acid	Sodium ascorbate, calcium ascorbate, Ascorbyl stearate and plamitate
BHA	
BHT	
TBHQ	
Erythorbic acid`	Sodium erythorbate
Gallic acid	Propyl gallate, octyl galalte, Dodecyl gallate
Gum guaiac	
Tocopherols	Vitamin E

The antioxidants most commonly used to stabilize food products against autooxidation are BHT, Mixture of 2-BHA and 3-BHA, n-propyl gallate or 3,4,5-trihydroxy benzoic acid propyl ester

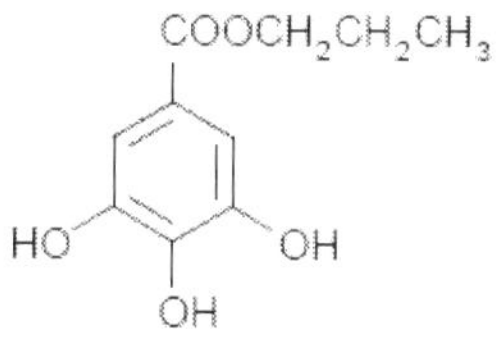

Kodak is marketing BHA and PG as Tenox BHA (Kodak) and Tenox PG (Kodak).

The above antioxidants are used in accordance with FDA regulations. in general. They are used to prevent oxidative degeneration of foods that affect flavor and odor. BHT is used in packaging materials for cookies, cold cereals etc. where high-vapor pressure of BHT permits its use. [33]

7.3 Stabilization of Fuels and Lubricants

Hydrocarbon fuels, Gasoline, Aviation turbine fuels, fuel oil, lubricating oils from petroleum sources - all need antioxidants to protect against auto-oxidation. The main phenolic antioxidants used in petroleum fuels are butylated phenols and cresols such as BHT, 2,4-dimethyl-6-tert-butyl phenol, and, 2,6-ditert-butyl phenols. Present use of antioxidant for petroleum fuels is in the order of 0.5 – 1 kg per 1000 barrels of fuel [1 US barrel = 159 liters].

For lubricating oils stabilization antioxidants used are 0.5 - 1.0% of zinc dialkyl dithiocarbamates in combination with 0.5 – 1.0% of BHT or 2,6-di-tert- butyl phenol. This is the recommended dose for mineral oil lubricants in gas turbines, steam turbine etc. Synthetic lubricants, both polyester and polyolefin types, are stabilized with 0.5-1.0% alkylated phenyl-α-naphthylamine and/or alkylated diphenylamine.

Commercial organic polymers, including elastomers, thermoplastics, synthetic fibers, and adhesives, are prone to oxidative degeneration during processing and end use. Antioxidants that influence performance includes compatibility or solubilities in the polymer or other substrate, volatility discoloration resistance, and thermal stability. Antioxidants are selected on the basis of these properties. Hindered phenolic antioxidants such as BHT, phosphites and non-staining amines are mostly used. [33]

7.4 Stabilization of Thermoplastics

Hindered phenols such as BHT, 4-methyl-6-tert-butyl phenols, 2,6-di-tert-butyl phenol, 2-tert-butyl-5-methyl phenol and other relative simple phenolic antioxidants have been traditionally used.

In recent years some Irganox™ types of antioxidants have been also preferred.

Hydrocarbon thermoplastic polymers, including polystyrene, polyolefins, and thermoplastic elastomers are stabilized with phenolic antioxidants, phosphites and phosphonites and thiosynergists, the composition and concentrations depend no life stage and end us as shown in Table 7.3. [33]

Table 7.3 Practices in Polymer Stabilization

Life stage of polymer	Stabilizer	Usual concentration, ppm
Drying	antioxidant	< 250
Storage	antioxidant	< 250
Compounding	antioxidant	500 –1000
	phosphite	500 – 1000
Fabrication	antioxidant	500 – 1000
	phosphite	500 – 1000
End use		
thermally	antioxidant	1000 – 5000
stressing	thiosynergist	1000 – 5000
Weathering	antioxidant	0 – 1000
	phosphite	500 – 1000
	hindered amine	0 – 1000
	UV absorber	0 – 10000

7.5 Antioxidants/Antiozonants for Rubbers

Rubbers degenerate more or less much under the action of heat, light, oxygen and ozone. By the action of heat, light and oxygen, free radicals are formed along the polymer chains; these radicals react either with oxygen or with adjaces rubber to form more free radicals and so on, resulting in a degraded polymer. Compounds capable of stopping these autocatalytic reactions are antioxidants.

Compounds, which aid protection against ozone attack, are called antiozonants. All antiozonants are also powerful antioxidants, though reverse is not true. Both raw natural rubber and raw synthetic rubbers already contain antioxidants. Natural rubber contains natural antioxidants, the antioxidants in synthetic rubbers are deliberately added during the manufacturing process in order to improve polymer stability. Still an extra amount of antioxidants, or antiozonats is nearly always added in order to render the ultimate vulcanizates more resistant to heat, light, oxygen or ozone. The added amount is 1-2 parts of antioxidant for 100 parts of rubber.

Substituted secondary and tertiary amines, substituted phenolic compounds, substituted phenol sulfides, organic phosphites etc. are the preferred antioxidants.

7.6 Antioxidants for Cosmetics

Many cosmetic preparations, particularly these containing animal and vegetable oils, may deteriorate rapidly due to oxidation. The spoilage may be substantially delayed by incorporating antioxidants. BHT, BHA, propyl gallate and Vitamin E are the most common antioxidants for cosmetics. [20] Important synthetic antioxidants having diversified applications in different products including food items have already been discussed. Natural antioxidants isolated from plants and animals are mostly used in food products.

Natural antioxidants are usually not used in products stabilization other than food. Firstly, because synthetic antioxidants are produced in bulk quantities (several thousand tons per annum) and are used in products, particularly petroleum and allied products, which are produced in, bulk quantities. Secondly, synthetic antioxidants are much cheaper than antioxidants derived from natural sources. Also from points of view of toxicity and carcinogenicity natural antioxidants are just suitable for natural products, such as food products where any addtive/preservative/antioxidants should be approved by FDA, WHO/FAO etc.

Synthetic antioxidants such as BHT/BHA, TBHQ, PG, are gradually being either banned or are restricted for food applications. In sum, out of the antioxidants produced from natural products. For other products synthetic antioxidants are preferred.

Though most of the antioxidants discussed earlier have been commercially exploited but one should keep in mind about the end uses. Though some other antioxidants have been detected or developed but yet not commercially exploited, they would call for additional data and information and only future R&D work will justify their commercialization or otherwise.

7.7 Synergism

It has been agreed upon on all hands that synergism can be expected between substances with different modes of action. Among natural substances synergistic effects are observed in the following [3]:

- Radical scavengers: tocopherols, spice extracts (carnosic acid) flavonoids. Carnosic acids generally are derived from rosemary extracts,
- Oxygen scavengers: ascorbates,
- Synergists (mostly acting as metal complexing agents, peroxide decomposers or reactivators of antioxidants; citric acid,

phosphoric acids, tartaric acid, lecithin, amino acids, ascorbic acid.

The synergistic effect permits the reduction of the level of antioxidants added to food formulations considerably. Reduction of cost mg/gm of the food product is same thing is true for synthetic antioxidants as the combination antioxidants of BHT-BHA, BHT-PG, BHA-TBHQ etc. These facts are true for both food products and other organic products including Rubber, Petroleum derivatives and compounds of crude oil after distillation or blending, plastics and of course fats and oils.

Important synthetic antioxidants having diversified applications in different products including food items have already been discussed. Natural antioxidants isolated from plants and animals are mostly used in food products. Natural antioxidants are usually not used in products stabilization other than food. Firstly, because synthetic antioxidants are produced in bulk quantities (several thousand tons per annum) and are used in products, particularly petroleum and allied products, which are produced in, bulk quantities. Secondly synthetic antioxidants are much cheaper than antioxidants derived from natural sources. Also from points of view toxicity and carcinogenicity natural antioxidants are just suitable for natural products, such as food products where any additive/preservative/antioxidants should be approved by FDA, WHO/FAO etc. Synthetic antioxidants such as BHT/BHA, TBHQ are gradually being banned or are restricted for food applications.

In sum out of the antioxidants produced from natural products, most of them are acceptable for food products. Though most of the antioxidants discussed earlier have been commercially exploited but one should keep in mind about the end uses.

Though some other antioxidants have been detected or developed but yet not commercially exploited, they would call for additional data and information and only future R&D work will justify their commercialization or otherwise.

8

CONCLUSIONS

8.1 SYNOPSIS

Preceding chapters have highlighted properties, chemistry and uses of antioxidants either derived from natural products of plant and animal origin or made synthetically.

Crude extracts of fruits, herbs, vegetables, cereals and other plant materials rich in phenolics are of increasing interest in the food industry because they retard oxidative degradation of lipids and thereby improve the quality and nutritional value of food. Scientists, manufacturers of food products and consumers are showing more interest in those antioxidants that are beneficial for prevention of coronary heart disease and cancer or other chromic diseases. Different expert in most of the countries are working for acceptance or rejection of a particular antioxidant for food products. FDA guidelines or recommendations of FAO/WHO accordingly are to be strictly compiled with. Great attention has to be paid for synthetic antioxidants as they are more susceptible for development of toxicity and carcinogenicity.

In case of petroleum products or petrochemicals or allied products like rubber, elastomers etc., however, such difficulties do not arise and synthetic antioxidants with proven track records are perfectly alright and recommended for use.

8.2 Current Scenarios

Selective solvents like acetone, methyl ethyl ketone (MEK), ethanol, methanol, n-hexane, methyl chloride, etc. have been successfully used for isolation of the active antioxidant ingredient from natural products. In some cases pure antioxidant (98-99% pure) product from the extracts has still not been recovered and attempts were made to reduce yield losses and improve the quality of the extracted antioxidant.

A relatively new process of extraction based on supercritical fluid extraction (SCFE) using high-pressure CO_2 (say at 350-400 bar and 70-80°C) has been suggested. SCFE process will not only help in improving the yield but also will guarantee the quality/purity of the final product. This process is no doubt a capital-cost intensive and calls for a detailed study of cost-benefit analysis.

R&D work carried out in a few reputed laboratories in the world including pilot plant studies at the Chemical Engineering Department of the Indian Institute of Technology (I.I.T.) at Mumbai, India have established the technology of SCFE extraction for quite a few products including antioxidants from natural sources. A few commercial plants using SCFE method have been established in some countries including India.

Hindered phenols and polyphenols were reported to have antioxidant properties in both antioxidants derived from natural sources or those made synthetically. However, it should also be emphasized that antioxidant activity does not necessarily correlate with high amounts of phenolics, and that is the reason why both phenolic content and antioxidant activity must be separately studied while establishing antioxidant potential for different substances. Sources for natural phenolic antioxidants are berries, apples, certain medicinal plants and vegetable peels, and different tree materials like root, bark, leaves etc. These have been discussed in appropriate chapters.

Propyl gallate, BHT, BHA, TBHQ etc., are established synthetic antioxidants and all of them are hindered phenolic substances. Most amazing antioxidants synthesized by Ciba-Geigy are Irganox™ group of antioxidants, which depending on the type and effectiveness, have covered almost any product, be it a food product, petroleum product or polymers derived from petrochemicals and natural or synthetic rubber or elastomers. Closely guarded through their own patents and proprietary rights Ciba-Geigy had the distinction of being the sole manufacturer of Irganox™ group of antioxidants.

However, patents have since expired and it is reliably learnt that one or two companies, involved in manufacture and marketing of other antioxidants, are planning to make Irganox™ type of antioxidants. A large number of antioxidants from natural sources were highlighted in Chapter 2. Among these antioxidants the five very important antioxidants, constituting an 'antioxidant network' as suggested by Dr. Packer and his co-workers and colleagues are the following:

Vitamin E
Vitamin C
Lipolic Acid
Glutathione
Coenzyme Q_{10}

These antioxidants were discussed in some details in Chapter 2. Chemistry, properties and their sources were highlighted and elaborated. It must be pointed out, as mentioned in the preceding chapters, both Vitamin C and Vitamin E are now synthetically made and both have become very high volume and extremely popular antioxidants.

Among synthetic antioxidants, which were discussed at some length in Chapter 3 and elsewhere, the more prominent and key antioxidants are the following:

BHT
BHA
Propyl gallate
TBHQ
Irganox™

Food additives including food antioxidants are subjected to pharmacological scrutiny and technical evaluation for mutagenic, carcinogenic and pathogenic effects. Such toxicity concerns and consumer's preference for 'all natural' ingredients and regulations discussed earlier have resulted in increased interest in natural antioxidants discussed in some details earlier.

Two very important tree extracts containing active ingredients in the form of antioxidants are:

Ginko biloba extracts and derivatives
Pycnogenol or pine tree extracts

These two groups of plant origin were highlighted in Chapter 2. These are flavonoids having multiplicity of uses against some chronic diseases are assuming increasing interest in many countries where these trees (Ginko trees and pine trees of special types) are not even grown but the extracts are imported for medicinal purpose.

Vitamin A, C and E are also assuming importance as natural antioxidants and recommended for disease prevention and cure.

Vitamin C is water-soluble
Vitamins A and E are fat soluble

These Vitamins are recommended not only for nutritional purpose but also as antioxidants for the human body as well as for other animals.

Many of the phytochemicals including flavonoids and carotenoids have been highlighted in Chapter 2.

Most prominent synthetic antioxidants including these mentioned above are butylated phenols, butylated cresols and also hindered polyphenols. These antioxidants including Irganox™ group of antioxidants have been discussed in details in Chapter 3.

Most important inorganic micronutrient acting almost like an antioxidant is Selenium. If the human body is suffering from inadequacy of Selenium there will be many problems. Along with selenium other less prominent micronutrients are zinc, magnesium, sulfur and also Manganese. All these inorganic elements are otherwise considered as essential for normal functioning of the body. So much so these have been termed by the author as semi-antioxidants.

8.3 Future Possibilities

If one thinks that a lot has been talked about antioxidants and very adequate data and information are available on both natural and synthetic antioxidants one will be wrong. Much more additional and authentic data are required for the antioxidants particularly with respect to safety, toxicity and carcinogenic effects of those antioxidants already commercially exploited and also those, which have been detected, devised, discovered and yet not commercially exploited.

During last fifty odd years a number of antioxidant laboratories and R&D centers have been established and very significant contribution is made by their scientists as a result of their investigations and findings.

There is nothing like a universal antioxidant. What was true two decades ago is no longer fully valid. As before, food antioxidants particularly synthetic ones, will call for very special re-investigations. Many valid questions have been asked regarding use of such versatile antioxidant like BHT, BHA and TBHQ regarding applications in food processing and food preservations. Many countries have either banned use of these antioxidants for food purpose or put them under restrictions. Reasons cited have been that from toxicity or carcinogenicity points of view these are not totally safe. For applications in other products other than food purpose however these are considered safe.

Some very important R&D works in a multipurpose antioxidants laboratory should be as follows:

i) Isolation and extraction of new antioxidants from natural sources in addition to those, which have been already covered.

ii) Role of butylation for production of butylated phenols and its homologues i.e. individual cresols. Why only butylation is being carried out? Alkylation through other olefins should be also established. Toxicity and carcinogenic effects on synthetic antioxidants should be studied in details and a data bank/data sheets for the individual butylated phenols and cresols should be established.

iii) Role of Selenium should be studied further. It is to be researched to reestablish the role of Selenium. Is there really something as inorganic antioxidant? Or, is it that they should always be called micronutrients like, Se, S, Zn, Mg or even Mn etc. What is the synergy with respect to these inorganic elements and established antioxidants?

iv) Are there other vitamins that can be also classified as antioxidants (other than Vitamins A, C and E).

v) Antioxidants from tree extracts need to be studied further. Just like Ginko and Pine trees there might be other trees whose extracts from bulk might be equally effective. This calls for more innovative and dedicated R&D work. India, China and some of the African countries might then other alternatives. All botanical aspects of some of these 'unknown' trees particularly in the forest areas need further work. More attention in this direction is definitely called for.

Let us hope that more and more natural antioxidants and of course new synthetic antioxidants will be discovered for the benefit of humanity and other animals.

vi) In case of antioxidants used for other products like petroleum and its many derivatives, rubber, plastics and others due attention has to be paid for green chemistry or eco-friendly antioxidants used as additives which indirectly should not affect human healths.

Once used as a universally additive for gasoline, TEL (tetra ethyl lead) has been almost totally phased out, as alternative technologies have been introduced which automatically augment the gasoline octane number and no TEL is added which is now considered as not eco-friendly and unsafe for human healths.

WHO/FAO and local regulatory bodies are very seriously working in these directions. Prime R&D organisations are also doing laudable work. Every research worker has to address seriously to the toxicity and carcinogenecity aspects of an antioxidants, what ever be their end uses. Protection of the total ecology and environment considering all aspects should be our collective work, and responsibility.

References

1. Lewis, J.R. (Sr.) in Hawley's Condensed Chemical Dictionary, 13th edition, Galgotia Publication Pvt. Ltd., New Delhi, 1998.

2. Lester, P., Carol, C. The Antioxidants Miracle, John Wiley & Sons, Inc. New York, 1999.

3. Hudson, B.J.F., In Food Antioxidants, Elsevier Science Publishers Ltd., Essex, England, 1990.

4. Merck Index, 13th edition, Merck & Co., Inc., New Jersey, 2001.

5. Journal of American Medical Association, 2005

   a) Jama ama-assn.org

   b) National Cancer Institute

      http://www.cancer.gov

   c) American Heart Association www.americanheart.org

6. Encyclopedia of Food Engineering, C.W. Hall, A.W. Farrache, A.L. Rippen, AVI Publishing Co. Inc., 2nd edition, West Point, Connecticut, U.S.A., 1986.

7. Holumn John R., Fundamentals of General, Organic and Biological Chemistry, 4th edition, John Wiley & Sons, New York, USA, 1990.

8. Debossy J.H., Materials and Technology, Vol.VII, Longman Group Ltd., Amsterm, The Netherlands, 1975.

9. Mukhopadhyay A.K., Industrial Chemical Cresols and Downstream Derivatives, Marcel Dekker, Madison Avenue, New York, USA, 2005.

10. Shah, C.S., In a textbook of Pharmacognosy, 9th edition, Shah Prakashan, R.S., Ed., Ahmedabad, India, 1992-93.

11. Feltman John, Healing Remedies & Techniques, The Berkley Publishing Group, 200 Madison Avene, New York, USA, 1996.

12. Ullman's Encyclopedia of Industrial Chemistry, 5th edition, Germany, 1996, Volume 38, pp. 218-224.

13. Vitamins, In Kirk-Othmer Encyclopedia of Chemical Technology, 4th edition, Wiley-Inter Science Publications, New York, 1994, Vol.25, pp. 178-180.

14. Karamta, H.K., M.Tech. thesis, 'Extraction and Enrichment of Natural Colors and Vitamins' , Chemical Engineering Department, I.I.T. Mumbai, 2004.

15. Setiawal, B., Sulaeman, A., Girand, D., Driskell, D., "Carotenoid Content of Selected Indonesian Fruits," Journal of Food Composition and Analysis 14, pp. 169-176, 2001.

16. Bhaskarachary, K., Sankarrao, D.S., Desthale, Y.G., Reddy, V., "Carotene Content of Some Common and less Familiar Foods of Plant Origin," Food Chemistry 54, pp. 189-193, 1995.

17. Astorg, P., "Food Carotenoids and Cancer Prevention" An Overview of Current Research" Trends in Food Science & Technology 8, pp. 406-413, 1997.

18. Website: www.genox.com/isa/isa-crotenoids.htm, 2001.

19. The Times of India, Mumbai edition, India, September 21, 2006, p. 14.

20. De bussy, J.H., Materials and Technology, Vol.V, Longman Group Ltd., Amsterdam, The Netherlands, 1972, p. 852.

21. Ash Michael and Ash Irene, Concise Encyclopedia of Industrial Chemical Additives, Edward Arnold Publishers, Kent TN 13, LYA, UK, 1991.

22. Handbook of Inorganic Chemicals, McGraw Hill, Professional Publishing, 2 Penn Plaza, New York, USA 2002, pp. 811-814.

23. Antioxidant activity of plant extracts containing phenolic compounds – Journal of Agric, Food Chem. 47,3954 – 3962, 1999 also published in 1999 by American Chemical Society on Web on 09/24/1999.

24.

i) Andrea G., European Patent No. WO 03079816, 2003

ii) Bernalte Gracia, M.J., Gonzlez, J.F., European Patent EP 1384472, 2004

iii) Ben-Yehuda, M., European Patent CN, 1198661, 1998.

iv) Guang, Hu., Xin, Li., European Patent CN 1298904, 2001

v) John, S., European Patent WO 0179355, 2001

vi) Morazzoni, P., Bombardelli, E., European Patent EP 0818225, 1998.

vii) Ruiz del Casrtillo Maria Luida, Blanch Monzano Gracia Patrica, European Patent WO 03103645

viii) Zwlkha Morris and Others, US Patent 5837311, 1998.

25. Md. Farak Ali, Bassam M., El Ali, Speight James G., Handbook of Industrial Chemistry (Organic Chemicals), McGraw Hill Companies, New York, USA, 2005 (pp. 115-118).

26. Harano, Y., Hoshino, O. and Ukita, T.J., Journal of Hyg. Chemistry (Japan), 13 (1967) p. 197.

27. Gallo-Tores, H.E. in Vitamin E ed., L.J. Machlin et al., Marcel Dekker, New York, 1980, pp 190-94.

28. Fujitani, T. and Ando, H., Japan Oil Chem. Society, 33(1984) pp. 354-56.

29. Kurechi, T., Aizawa, M. and Munugi, A., J. American Oil Chemical Society, 60 (1983) 1870-78.

30. Kurechi, T. and Kunugi, A., J. American Oil Chemicals Society, 60 (1983), 33 and 1892, pp.

31. Haigh, R., Food Chemicals Toxicology, 24 (1986), 1031.

32. FAO / WHO Codex Alimentarius Commission, Alinorm 87/17, Report on 13th session of the Codex Committee on Fats and Oils, London, 23-27 February, 1987, p. 5

33. Klemchuk, Peter P., Ullman's Encyclopedia of Industrial Chemistry, Wiley Tech-VCH GmbH & Co Germany, Vol.3, pp. 643-46, 1996.

34. Pauling L., Proceedings of National Academy of Science, USA, 68, 2678, 1971.

35. Kent James, A., Roegel's handbook of Industrial Chemistry, Kluwer Academic/Plenum Publishers, New York, 2003, Pages 1007-1008, pp. 256-258.

36. Milne, G.W.A., Gardner's Chemical Synonyms and Trade names, Eleventh edition, Gower Publishing Limited, 1999, U.K.

Glossary

ALZHEIMER'S DISEASE

After heart disease, cancer and strokes, it is the most common cause of death in the Western world. Attacking the brain's 'gray matter' is a disease of mental processes rather than physical function, characterized by memory loss and progressive intellectual impartment. It was first described by Alois Alzheimer in 1906.

ANTIOXIDANT

A group of organic chemical compounds responsible for inhibition of degradation of products encompassing food, fats and oils, rubber, plastics and other petroleum products. They are mostly of natural origin i.e., obtained from plants and animals. Also many of them are obtained through organic chemical synthesis.

ANTIOZONANTS

Ozone, being an extremely powerful oxidizing agent, readily oxidizes a variety of functional groups in biochemicals and other chemicals, such as rubber. Any chemical that inhibits the process of oxidation by ozone is called an antiozonant.

ASCORBIC ACID

Also known as Vitamin C is widely distributed in nature both in plants and animals and is a known antioxidant. Primarily isolated from citrous fruits. Commercially obtained by chemical synthesis mostly from D-glucose.

BHA

Butylated hydroxy anisole has become a very popular anti-oxidant particularly for feed and food products. Made by the reaction of p-methoxy phenol and isobutylene.

BHT

Butylated hydroxy toluene or di-isobutyl-para cresol (DBPC) is one of the most prominent synthetic antioxidant made from p-cresol and isobutylene. Widely used as an antioxidant to inhibit oxidation for a range of products including food, feed, petroleum products, plastics and polymer, rubber, fats and oils etc.

CAROTENOIDES

Colored plants chemicals or terpenes, which are oil soluble include yellow orange, β-carotein, a precursor of Vitamin A, xanthophylls, yellow pigment lutein in plant leaves, and Zeaxanthin in corn, lycopene the red coloring of tomatoes etc. The carotenoids are generally highly susceptible to oxidation. Only selected members (Xanthophylls, β-carotene and bixin) are extracted for food coloring, used as antioxidants.

CITRIC ACID

A very popular food additive and antioxidant, is widely distributed in plants and in animal tissues and fluids. Fermentation of sugar/molasses using Aspergillus niger strains produces citric acid industrially known as acidulant in beverages, confectionary, in pharmaceutical syrups etc.

ERYTHORIBIC ACID

Also known as isoascorbic acid, which does not occur naturally. Its biological activity is only 5% of ascorbic acid. In most of the European countries it is not permitted as a food additives.

FLAVONOIDS

A group of aromatic, oxygen containing, heterocyclic pigments widely distributed among higher plants. They constitute most of the yellow, red and blue colors in flowers and fruits. Not to be confused with carotenoids.

GINKO BILOBA

The Ginko is a strong, hardy tree of the Stone Age. It managed to survive in China and its extracts were found to have medicinal and antioxidant properties. Transplanted to USA, it is now commonly grown in parts of USA and Europe apart from China.

GRAS

Abbreviation of, generally regarded as safe. From toxicity and carcinogenicity points of view food additives/antioxidants should have GRAS qualities.

HUNDRED PHENOLS

A major class of synthetic antioxidants which have been used for inhibition of oxidative degeneration of various organic products such as plastics, rubber, feeds, synthetic fibers, fuels, lubricants, etc.

IONOX^{TM}

A group of trade mark antioxidant chemicals manufactured and marketed by Shell, UK. IONOX-100 and IONOX-210 are examples of these speciality antioxidants.

IRGANOX^{TM}

A group of highly effective trade mark antioxidants introduced by Ciba-Geigy some two decades ago. There are a number of such antioxidants designated by separate numbers. These groups of antioxidants cater to almost all organic products. In fact they were engineered to act as antioxidants for various products that included food, feed, polymers and plastics, rubber etc.

PARKINSON'S DISEASE

Degenerative disease of the brain characterized by a progressive loss of mobility, muscular rigidity, tremor and speech difficulties. The condition is mainly seen in people over the age of 50.

PHYTOCHEMICALS

Natural compounds found in fruits and vegetables. Specifically important in plant growth and metabolism and plant products. Many of them are antioxidants.

PHYTOL

An alcohol obtained by the decomposition of chlorophyll. One very important use is in the synthesis of Vitamins E and K.

PROVITAMIN

The precursor of a vitamin, carotenes (particularly Beta-form) are the best examples. Cartoene itself is not a vitamin, however, upon activation in the system (ingested in the liver) it is converted to Vitamin A.

PYCNOGENOL

An extract from the bark of a French Maritime Pine tree, has been used extensively for medicinal purposes. Recently re-established as a powerful antioxidant.

RDA

Abbreviation of, Recommended Daily (or Dietary), Allowances for intakes of food products as established by the Food and Nutrition Board of the National Research Council, National Academy of Sciences. Mostly used for Vitamins, minerals (micronutrients) and other food items.

RETINOL

A component of Vitamin A. Fat soluble chemical derived from β-carotene and found in milk, butter, cheese, egg yolk, liver etc. Lack of retinol in the diet leads to the eye disease xerophthalmia.

SCAVENGERS

or, more appropriately termed as oxygen scavengers such as ascorbic acid (Vitamin C), ascorbyl palmitate, erythorbic acid and its sodium salt, which react with oxygen and can thus remove it in a closed system.

SYNERGISM

In many cases it was observed that a few antioxidants as a combination was more effective than individual antioxidants. This was evidently as a result of synergistic effects. For instance BHT-BHA in a predetermined ratio was more effective as a combination antioxidant than either BH T or BHA alone.

TRACE/ELEMENT

Any element such as Mn, Zn, Mg, etc., that the body needs each day in an amount of no more than 20 mg.

VITAMIN A

Retinol, a fat-soluble vitamin in yellow colored foods and needed to prevent night blindness.

VITAMIN E

A mixture of tocopherols, a fat-soluble vitamin apparently needed for protection of edema and anemia (in infants) and possibly against dystrophy, paralysis, and heart attacks.

Made in United States
Orlando, FL
04 November 2026

90200480R00095